Lecture Notes in Earth Sciences

Lecture Notes in Earth Sciences

Edited by Somdev Bhattacharji, Gerald M. Friedman,
Horst J. Neugebauer and Adolf Seilacher

29

Fritz K. Brunner   Chris Rizos (Eds.)

Developments in Four-Dimensional Geodesy

Selected papers of the
Ron S. Mather Symposium on Four-Dimensional Geodesy
Sydney, Australia, March 28–31, 1989

Springer-Verlag
Berlin Heidelberg GmbH

Editors

Fritz K. Brunner
Chris Rizos
School of Surveying, University of New South Wales
P.O. Box 1, Kensington NSW 2033, Australia

ISBN 978-3-540-52332-1   ISBN 978-3-540-46961-2 (eBook)
DOI 10.1007/978-3-540-46961-2

© Springer-Verlag Berlin Heidelberg 1990
Originally published by Springer-Verlag Berlin Heidelberg New York in 1990

2132/3140-543210 – Printed on acid-free paper

PREFACE

This monograph is a compendium of revised papers which were originally presented at the "Ron Mather Symposium on Four-Dimensional Geodesy", 28-31 March, 1989, held at the University of New South Wales, Sydney, Australia. The symposium had the enthusiastic support of the International Association of Geodesy and the Australian Academy of Sciences. The symposium served two purposes: to honour the achievements of the late Professor Ron S. Mather, the distinguished Australian geodesist who died in 1978, and to review and report on the latest developments in four-dimensional geodesy.

Four-dimensional geodesy is a convenient term for those geodetic principles and techniques which yield position, gravity and their time variations. In the past geodesists have tended to think of the earth as a static body, save from occasional savage earthquakes or volcanic eruptions. So, why the need to coin the term "four-dimensional geodesy"? Because it explicitly recognises that time is an integral part of understanding geodetic measurements. But let's first identify the scope of modern geodesy. Geodesy has traditionally been concerned with two separate, though closely related, topics: accurate positioning of objects on the earth's surface, and mapping the earth's external gravity field. These are still the fundamental tasks of geodesy, although the spheres of application have now extended into space. However, present and emerging geodetic measurement technologies for gravity field mapping and positioning are sensitive to deformations of the earth's surface and gravity field.

Within the geodetic community, this new emphasis on accounting for the time-varying characteristics of position and gravity has fundamental principles; in particular the establishment and maintenance of appropriate global reference systems for geodesy. At the same time, there has been a growing recognition by the earth sciences in general of the important role of geodesy in studying earth deformations, as well as atmosphere and ocean dynamic phenomena. The geodetic measurements, for example, are taken over time scales of hours to decades, and occasionally to a century or longer. Though this is only a small part of the whole deformation spectrum, it is a very important one. Geodesy bridges the low frequency part of the spectrum available from geological observations, with the high frequency end observed from, for example, seismic instrumentation. It's role in atmospheric and oceanographic studies is as a unique, high precision remote sensing tool.

The revolution in geodesy is not, however, restricted to the measurement technology only. It is true that without the advances of space geodesy and terrestrial metrology, the notion of four-dimensional geodesy is a rather academic one. These advances, which now reveal time-variable signals above the measurement noise level, have important implications for all geodetic activities. The geodetic activities we refer to can be identified as: experiment design and measurement processes; definition and maintenance of highly stable geodetic reference systems; data analysis; and interpretation of position and gravity results. Ultra high precision measurements are of little use without sophisticated analysis tools to extract the small signals in the data. The interpretation of geodetic results will be in error if insufficient attention is paid to ensuring that the reference systems to which the results relate are themselves stable. Clearly four-dimensional geodesy is as much about concepts and principles, as about computers and geodetic equipment.

This diversity is reflected in the papers selected for this book. They range over topics related to the modern measurement tools, the reduction and analysis techniques, to the interpretation of geodetic results within the context of problems currently being investigated in the earth sciences.

We would like to thank the International Association of Geodesy and the Australian Academy of Sciences for sponsorship of the Symposium. Unisearch Ltd., the commercial arm of the University of New South Wales, was the managing agent, and staff members of the School of Surveying and of Unisearch Ltd. were involved in the organisation of the Symposium. We would like to gratefully acknowledge these excellent contributions.

Let us express also our gratitude for the useful guidance which we received from Prof. K. Lambeck, A. Prof. A. Stolz and Dr. R. Coleman of the Scientific Advisory Committee and the continuous support given by Prof. E.W. Grafarend.

Sincere thanks are due to the authors of the selected papers for agreeing to contribute to this Monograph, and for their positive cooperation during the production of this volume.

November 1989
Sydney, Australia

Fritz K. Brunner
Chris Rizos

CONTRIBUTORS

BRETTERBAUER, Kurt
        Institute of Theoretical Geodesy and Geophysics
        T.U. Vienna, Gusshausstrasse 27-29,
        A-1040 Vienna, Austria.

CHEN, Ruizhi
        Finnish Geodetic Institute,
        Ilmalankatu 1A,
        JF - 00240 Helsinki, Finland.

FORSBERG, Rene
        National Survey and Cadastre,
        Gamlehave Allee 22,
        DK-2920 Charlottenlund, Denmark.

GRANT, Donald B.
        Department of Survey & Land Information,
        Head Office, P.O. Box 170,
        Wellington, New Zealand.

GROTEN, Erwin
        Institute of Physical Geodesy,
        TH Darmstadt,
        Petersenstrasse 13,
        6100 Darmstadt, Federal Republic of Germany.

HECK, Bernd
        Department of Geodetic Science,
        University of Stuttgart,
        Keplerstrasse 11,
        D-7000 Stuttgart 1, Federal Republic of Germany.

ILK, Karl-Heinz
        Institute for Astronomical and Physical Geodesy,
        Technical University of Munich,
        Arcisstrasse 21,
        D-8000 Munich 2, Federal Republic of Germany.

KAKKURI, Juhani
        Finnish Geodetic Institute,
        Ilmalankatu 1A,
        JF - 002040 Helsinki, Finland.

KEARSLEY, William A.H.
School of Surveying,
University of New South Wales, P.O. Box 1
Kensington, NSW 2033, Australia.

KLEUSBERG, Alfred
Geodetic Research Laboratory,
University of New Brunswick,
P.O. Box 4400,
Fredericton, NB, Canada E3B 5A3.

KOLENKIEWICZ, R.
NASA Goddard Space Flight Center,
Laboratory for Terrestrial Physics,
Greenbelt, Maryland 20771, U.S.A.

LAMBECK, Kurt
Research School of Earth Sciences,
The Australian National University, GPO Box 4,
Canberra, ACT 2601, Australia.

REILLY, W. Ian
Geophysics Division,
Department of Scientific and Industrial Research, P.O. Box 1320,
Wellington, New Zealand.

SMITH, D.E.
NASA Goddard Space Flight Center,
Laboratory for Terrestrial Physics,
Greenbelt, Maryland 20771, U.S.A.

TEUNISSEN, Peter J.G.
Geodetic Computing Centre,
Delft University of Technology, Thijsseweg 11,
2629 JA Delft, The Netherlands.

TORGE, Wolfgang
Institute of Geodesy,
University of Hannover,
Nienburgerstrasse 9,
D-3000 Hannover 1, Federal Republic of Germany.

STOLZ, Artur
School of Surveying,
University of New South Wales, P.O. Box 1
Kensington, NSW 2033, Australia.

TABLE OF CONTENTS

The Fourth Dimension in Geodesy:
Observing the Deformation of the Earth

K. Lambeck
Research School of Earth Sciences
Australian National University
Canberra, Australia

ABSTRACT:

The earth is a complex body that deforms over a wide range of length and times scales. Observation of these deformations constrain models of the unknown forces (e.g., plate tectonics driving forces) and models of the planet's response to known forces (e.g., tidal or rotational forces). Geodetic measurements, in particular those based on the ultra high precision space-age technologies, are central to the study of these deformations. The geodetic measurements cover time scales of hours to decades and occasionally to a century or longer. This is only a small part of the whole deformation spectrum. Other parts are available from geological and geomorphological observations (at the low frequency end) and from seismic instrumentation (at the high frequency end). The geodetic data provides an important bridging of these other data types. They will elucidate known phenomena that presently only rise marginally above the noise levels of existing methodologies and new signals will appear. When combined with new developments occurring in other areas of the earth sciences the geodetic methodologies will contribute significantly to our understanding of the working of the earth.

1. INTRODUCTION

The title of four dimensional geodesy for this conference recognises that the time element is an integral part of understanding geodetic measurements. In the past geodesists have tended to think of the Earth as a static body, occasionally distorted by earthquakes or its surface punctured by volcanic eruptions. This view is largely understandable because the time scale of the more global and obvious deformations have been much longer than our own life spans. But when the Earth is viewed on geological time scales we see a very different story. Far from being static, we see a planet that is rent asunder at ocean ridges. We see the buckling of continents in collision and crust being recycled back into the mantle. We see islands and mountains rising out of the sea and large segments of crust subsiding. Once the time scale is collapsed we see a very dynamic Earth indeed. But even on the human time scale the planet is indeed an active entity once it is put under the microscope of modern geodetic measurements. Tidal and rotational deformations occur with periods of hours to years. Global deformations of the planet occur on a variety of time scales in response to changing surface loads in the atmosphere, oceans and hydrosphere.

With the methods of space geodesy now available a large part of the spectrum of these deformations has risen above the measurement noise level but another large part still remains inaccessible because of the very long time scales involved. This part of the record remains locked up in geological observations and one of the challenges of modern geodesy is to integrate this part of the spectrum of the Earth's deformation with that part established by geological observations and, at even higher frequencies, with that part of the spectrum explored by seismologists. (Geodesy can be seen merely as high frequency geology or as low frequency seismology.) The challenge is to establish the links with records contained in rocks such as the one illustrated in Figure 1a with the observations derived from the radio telescope illustrated in Figure 1b. In this case the two span the extreme ends of the spectrum of the Earth's deformations. This particular rock from Western Australia contains 4.2 billion year old minerals, the oldest known terrestrial material. It indicates that crust was already being created and destroyed at that time. Between 2.67 and 3.1 billion years ago the sediments hosting these minerals were deposited and buried to a depth of more than 15 km and subjected to temperatures in excess of 500°C. Later it found its way back to the surface where it has remained for perhaps the past billion years. In comparison, the deformations recorded by the radio telescope (Figure 1b) at the other side of the world represent only a miniscule fraction of the Earth's history.

(a)

(b)

Figure 1: Two recorders of Earth Deformation.
(a) A conglomerate from the Jack Hills area of Western Asutralia which contains a record of Earth deformation spanning 4.2 billion years.
(b) A radio telescope used for long baseline interferometric observations of Earth deformation on time scales of hours to years.

2. GEODESY AND THE PLATE TECTONICS HYPOTHESIS

The plate tectonics hypothesis has provided a marvellous synthesis of much of the dynamic behaviour of the Earth for the past 10% of the planet's history. In the present climate when questions of relevance are constantly being raised it should not be overlooked that the hypothesis has been more than simply an exciting scientific development. It has also led to an understanding of mineralization processes and hydrocarbon accumulation that are leading to new resource discoveries in a number of different tectonic settings. Also important, at least for a small segment of society, is that the hypothesis has given a new lease of life to the subject of geodesy. With the high accuracy instrumentation that is now available there is simply no place for static Earth concepts. The planet must be seen as a deformable body over a wide range of time scales. This is well recognized by this conference with its emphasis on the fourth dimension.

The hypothesis is essentially a kinematic one in which average motions of large tectonic units occur, one relative to another. What permits the motion to occur is largely a matter of describing what happens at the boundaries between adjacent plates. Cartoons of subduction tectonics and of ocean ridge spreading (Figure 2) are familiar parts of the Earth science literature but what is less well understood is the quantification of the process involved. We need to know the forces operating and we need to know the rheology of the Earth; how it responds to these forces. Here geodesy plays an important role.

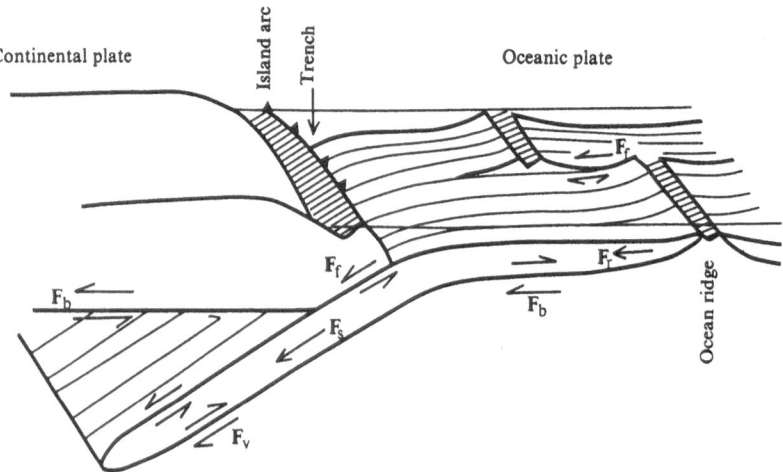

Figure 2: Cartoon of some of the tectonic processes occurring at plate margins. New crust forms at the ocean ridge to be subducted back into the mantle at a later date. The various forces F operating are understood largely in qualitative terms only.

One axiom of the plate tectonics hypothesis is that the plate motions are uniform on time scales of a million years or longer but this may be an artifact of the resolution of the geological observations. What is required is high temporal resolution of the plate motions and this is an obvious role for geodesy. A number of recent geodetic experiments are showing that the present-day motions are very similar to average motions for the past few million years, (e.g. Smith et al. 1989; Stolz et al. 1989). The implications of this are important for it suggests a sufficiently tight coupling between lithosphere and asthenosphere for the plates not to respond episodically and abruptly to changes in stress at the plate margins.

The inter-plate motions can be expressed in a number of ways, as baseline expansion rates, for example, or as relative rotation rates of the plates. The latter are particularly useful because they are independent of estimates of the height component of the stations, generally the least well determined coordinate. Furthermore, they permit straight-forward comparisons to be made with the geological estimates for the rotation vectors of the plates. Table 1, from Lambeck (1989), illustrates results based on the baseline expansion rates between the Australian, Pacific and Eurasian plates of Smith et al. (1989) and the results are essentially in agreement with the geological estimates of Minster and Jordan (1978).

Another axiom is that the plate boundaries, usually drawn as simple lines on maps, are sharply defined and that all inter-plate motion occurs on these boundaries. Closer inspection of the geology or seismic evidence indicates that more often than not these deformations occur over a wide zone and the line on the map turns into a complex zone of up to 500 km or more wide. Here the geodetic observations again play a role, in this case in defining how the motions between adjacent plates are absorbed; in defining the strain field across the boundary from which the stress field can be deduced if the rheology is known.

Conventional geodetic measurements have been important here. Much of what we know about the stress-strain cycle at plate boundaries of the transform type was, for example, already elucidated early this century thanks to geodetic measurements made on the San Andreas Fault of California (NOAA, 1973). Considerable insight into the stress-strain cycle at subduction type convergent margins has been derived from early geodetic observations in Japan (Tsuboi, 1933). Particularly illustrative have been the geodetic observations of the past century for New Zealand because of the way in which the geodetic displacements have been transformed into strain and relative velocities that can be compared directly with the palaeomagnetic evidence for plate motions (Walcott, 1984).

Table 1: Baseline expansion rates ds/dt (from Smith et al. 1989) and relative rates of rotations estimated from individual baselines, and relative rotation rates from Minster & Jordan (1978)

Baseline	ds/dt (mm/a)	$\Omega(°/Ma)$	$\sigma_\Omega(°/Ma)$
Australian-Pacific plate			
Orroral-Hawaii	-77±4	1.407	0.073
Orroral-Huahine	-86±4	1.638	0.076
Yaragadee-Hawaii	-89±2	1.077	0.024
Yaragadee-Huahine	-78±4	1.397	0.072
Mean		1.173	0.065
Minster & Jordan (1978)		1.25	0.02
Pacific-Eurasian plate			
Hawaii-Simosato	-68±4	-0.662	0.039
Huahine-Simosato	-78±7	-0.720	0.065
Mean		-0.677	0.045
Minster & Jordan (1978)		-0.98	0.03
Eurasian-Australian plate			
Simosato-Orroral	-55±5	-0.573	0.052
Simosato-Yaragadee	-62±3	-0.555	0.027
Mean		-0.558	0.041
Minster & Jordan (1978)		-0.70	0 02

What these studies have emphasized and what is equally important for the new class of space technology based instrumentation, is:

(i)   The strain fields across plate boundaries are considerably more significant than displacements between isolated points on adjacent plates. A high density of points is required across the margin in order to establish the strain field.

(ii)   Short series of high-precision observations are no substitute for long series of observations repeated frequently. In some instances a high frequency of repeat observations may actually be more important than very high precision, although the latter is of course always desirable. New surveys, particularly with GPS, will therefore be much enhanced if they are built on older geodetic networks.

(iii)   Integration with geological and geophysical data is essential.

A further axiom is that the plates, away from their boundaries, behave essentially as rigid bodies, moving over the globe relative to each other without undergoing distortion. It would be truly remarkable if the irregular shaped plates, acted on by a variety of forces along its boundaries, can move relative to each other over an ellipsoidally shaped surface without undergoing some internal deformation. What this axiom implies, therefore, is that either these deformations are small compared with the motions at the plate boundaries or that these internal distortions are very small when averaged over intervals of millions of years. That the plates undergo some internal deformations can be seen in the seismicity that occurs within plates well away from known plate boundaries. The Australian continent, generally believed to be tectonically stable, has been subject to significant seismic activity ever since monitoring began (Figure 3). What is required is high temporal resolution of the plate motions. Clearly this is a role for geodesy. A number of recent studies are showing that the internal deformations, if occurring, are smaller than the intraplate motions but it remains important that this axiom is continually tested in any experiment for measuring inter-plate motions, if for no other reason that is provides a test of the validity of the geodetic experiment. The Smith et al. (1989) solution, for example, gives non-zero baseline expansion rates for a number of intraplate baselines but it would be premature to conclude that plate deformation occurs.

Important in the geodetic studies of plate tectonics is the measurement of vertical movement. With the emphasis placed on the horizontal displacements there has been a tendency to neglect the vertical component. This is understandable for not only is this latter component much smaller, it also does not exhibit the simple global patterns exhibited by the horizontal displacements. Nevertheless, they are an essential ingredient in the study of the Earth's deformation. In particular, vertical movements are often manifestations of horizontal forces at work and major uplifts are possible. Spectacular examples include the Huon Peninsula of Papua New Guinea where uplifts of 400 m in as little as 100000 years have occurred in response to the compressional interactions between the Australian and Pacific plates.

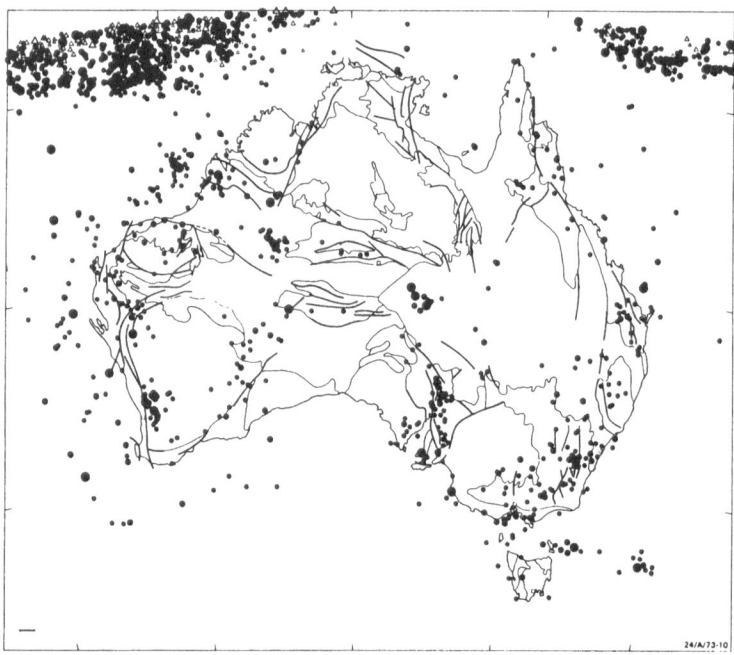

Figure 3:   Map of Australian seismicity of events of magnitude 4 or greater recorded from 1873-1980. The seismicity to the north defines the northern boundary of the Australian plate.   AG refers to the Late Proterozoic-Cambrian Adelaide Geosyncline and LFB refers to the Palaeozoic Lachlan Fold Belt. Two zones of intraplate deformation suggested by Cleary and Simpson (1971) are indicated by the dashed lines.

3. GEODESY AND THE HIGH FREQUENCY DEFORMATIONS OF THE EARTH

Plate tectonics does not provide the sole rationale for developing the geodetic discipline. In particular, the planet undergoes a number of deformations at periods shorter than the geological time scale whose closer investigation is of intrinsic interest as well as of relevance to understanding the workings of the planet on the longer time scale.

To understand the workings of the Earth requires a knowledge of the forces acting on the Earth and of the response of the planet to these forces. In some instances the forces are well known, such as the tide raising gravitational potential or the centrifugal force. Here the observations of the response of the planet establishes a stress-strain relation whose proportionality constants define the rheology of the planet appropriate for this particular problem. In a second class of problems deformations are observed but the forces are largely unknown. One example of this is the nature of coupling of core motions to the mantle and vice-versa. Here a rotational response is observed and attributed to such a general mechanism but whether this coupling is electromagnetic, viscous, or topographic remains largely a matter of choice (Lambeck, 1980). The former class of problems, of studying the stress-strain relation, is of considerable importance. In the case of the tidal deformations, for example, the rheological constants are usually expressed as Love numbers and phase lags or attenuation factors, and the central problem is to obtain representative observations of small deformations over the tidal spectrum. Because of attenuation of stress cycles both the Love number magnitude and the phase lag are expected to be frequency dependent and the objective is to measure their dependency over the tidal band from 12 hours to 18.6 years. This task is not easy. Ocean tides contaminate the results and over the longer periods tectonic deformation may mask the tidal signals. Models for the fluid tides need to be improved but this in itself requires that the Earth's response to surface loading be known. The two types of tidal deformation - solid and fluid - are inextricably linked. Yet progress in this area is desirable for it will improve both solid Earth and ocean understanding. The longer period tidal deformations are also contaminated by meteorological signals, including the loading of the Earth's surface, ocean and land, and this needs to be taken into account as well.

The waxing and waning of the ice sheets provides another example of quasi-periodic forcing of the Earth. The Late Pleistocene collapse of the ice sheets and the addition of water into the oceans results in a redistribution of the surface loads on the Earth. The result is crustal subsidence where the water load is increased and crustal rebound where the ice sheet has vanished. Globally, the flow induced in the mantle changes the inertia tensor and gravity field of the planet with the concomitant changes in rotation. The glacial rebound problem differs from the tidal problem in several ways. First, the characteristic period of the former is of the order of 10000 years and only the tail-end of the last cycle of deformation can be observed by geodetic techniques. Second, the load has a much greater spatial variation than the tidal force and the Earth's response contains, in consequence, a correspondingly greater amount of information of the Earth's rheology. Third, the load is only partly known and the further back one goes in time the more poorly it is known and the more uncertain become the estimates of the response. Geodetic observations alone do not suffice to resolve this problem: geomorphological evidence of past vertical movements of the crust relative to sea-level provide an essential data set and glaciological evidence and arguments are an important input into the reconstruction of the load function.

At longer periods another example of surface loading problems is provided by the loading of the crust by large volcanic structures, particularly in oceanic environments. Here the loading occurs on the time scales of 10^6 years and the response is measured in terms of the net displacements of the crust, either directly by measuring the shape of the sea-floor topography or by geomorphological evidence of the uplift of surrounding islands, or inferentially by measuring gravity or geoid height. Here the load is almost as great an unknown as the response and the supplementary observations are from seismology, geomorphology, geochronology and geology.

What these various observations permit us to establish, at least in principle, is a spectrum of Earth deformations from which the rheology function can be established. The range of relevant processes are illustrated in Figure 4. The rheology function will not be simple. Firstly, it will exhibit some depth dependence with high strength for the lithosphere and low viscosity for the asthenosphere. Secondly, the function will exhibit frequency dependence. At the seismic end of the spectrum the mantle responds primarily as if elastic and anelastic effects are secondary, but at very long periods, corresponding to seamount loading for example, the mantle behaves essentially as a fluid. How the function varies at the intermediate frequencies remains unclear and a worthy objective for geodetic studies. Thirdly, the function may also exhibit stress magnitude dependence, with the planet responding faster to large loads than to small loads. Fourthly, the function will certainly exhibit lateral variations for there is abundant geophysical evidence for lateral variations in a variety of physical properties of the Earth. Once this function, or parts of it, is mapped it becomes possible to make predictions about the mechanical forces responsible for the other deformations. It becomes possible, for example, to draw conclusions about mantle convection and the driving forces of plate tectonics. If, for example, the mantle viscosity increases significantly with depth then convection may be largely restricted to the upper mantle. Little mixing with the lower mantle may result and lead to different chemical and isotopic signatures of the volcanism at ocean islands and mid-plate hotspots. If the mantle viscosity is more uniform then a greater degree of mixing of the upper and lower regions of the mantle may occur and the chemical composition is likely to be more homogeneous. Clearly any geodetic observations that lead to improved mantle viscosity estimates make an important contribution in constraining models of the Earth's evolution.

What geodetic observations are important here? Global gravity field or geoid height measurements are one obvious answer. The "secular" part of the field constitutes a measure of the response of the Earth to the very long period forces associated with plate tectonics and mantle convection. There would be little dispute these days with the argument that this field reflects the dynamics of the Earth's mantle on time scales of 10^{6}-10^{8} years and that it constrains, in principle at least, models of mantle convection and plate driving forces. After all, convection in its simplest definition is the motion resulting from the gravitational forces acting on lateral and radial density variations. But just how to use these observations most effectively remains a difficult matter because of the fundamental non-uniqueness of interpreting gravity fields. Complementary geophysical data are required and the most exacting ones come from the methods of seismic tomography by which the three dimensional structure of the mantle is being mapped (e.g. Dziewonski, 1984; Woodhouse et al., 1984). But much progress needs to be made in this discipline before we have a spatial resolution that begins to approach the resolution attainable with gravity or geoid observations. Nevertheless, the combination of the gravity and seismic data is beginning to provide important new insights into the mantle viscosity (Richards et al., 1984).

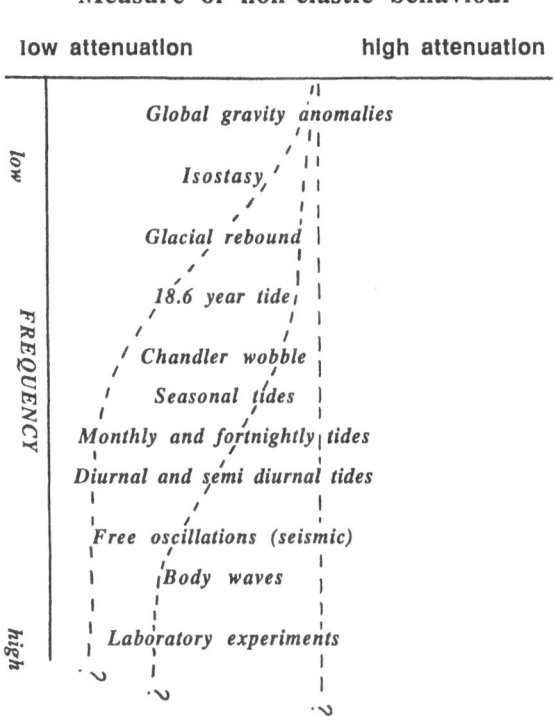

Measure of non-elastic behaviour

low attenuation   high attenuation

Global gravity anomalies

Isostasy

Glacial rebound

18.6 year tide

Chandler wobble

Seasonal tides

Monthly and fortnightly tides

Diurnal and semi diurnal tides

Free oscillations (seismic)

Body waves

Laboratory experiments

FREQUENCY   low   high

Figure 4:   Spectrum of Earth-deforming processes. The rheology function is not (cannot be) defined. Several schematic examples of how this function may vary with frequency are shown.

The gravity observations play another important role in understanding the long term dynamics of the mantle through the study of lithospheric structure. The altimeter satellites have provided an unprecedented high resolution image of the gravity field over the oceans (e.g. Haxby et al., 1983) but comparable resolution images over the continents await a new generation of satellites or the rapid opening up of national borders to terrestrial gravity surveys. The altimetry data, when combined with seismic and other geophysical and geological observations, has led to the understanding of the evolution of the mechanical properties of the ocean lithosphere and provides constraints on the boundary conditions that this layer imposes on mantle convection. The altimeter satellites have also led to the identification of numerous new features in the ocean floor and have provided one of the best ways to provide an approximate but quick survey of the ocean floor topography.

The time dependence of the gravity field has been an important subject of study for many decades through the measurements of the tidal deformations. Important developments have been the high precision absolute gravity meters and the highly stable cryogenic gravimeters for measuring the terrestrial deformations out of the seismic frequency band. Of considerable significance are the measurements of the long period tides and the rotational tide, or pole tide. One parameter of interest is the lag in the response of the Earth to the tide raising potential although very few, if any, significant measurements yet exist, in part because of unknown instrumental lags and in part because of the oceanographic perturbations of the solid tide signal. The ocean-solid tide interaction remains a problem and a close interaction with physical oceanography is essential. The other parameter of importance is the amplitude of the tidal response, particularly the frequency dependence of the amplitude across the diurnal band of the spectrum because of the core resonance phenomenon (Wahr, 1981). The amplitude variation over the longer periods resulting from the planet's departure from elasticity have the potential of measuring the non-elastic response over a frequency range from hours to years but here also the results are perturbed by ocean tides and, at the seasonal frequencies, by meteorological factors.

The terrestrial measurements of the tidal response is contained in Love number combinations of the form $(1+k_n-k_n)$ or $(1+2h_n/n-k_n(n+1)/n$ with most observations being limited to $n=2$. These functions are less sensitive to the frequency dependent processes than the individual Love numbers themselves and an important development of the past decade has been the ability to measure the potential Love number k_n alone from the analyses of satellite orbits. Other than this response, the displacement Love numbers h_n, l_n are worth investigating more closely because they reflect more regional and even local responses and because they provide independent measures of the planet's elastic and anelastic parameters.

Also important are the high precision analyses of the LAGEOS satellite orbit for time-dependence of the gravity field through the measurement of the time dependence of the zonal stokes coefficients J_n. Recent results by Cheng et al. (1989) are particularly interesting.

4. CONCLUSION

From the few examples raised here it is clear that the Earth is a very dynamic planet in whose study the geodetic measurements are playing an ever increasing role. The essential characteristic of the geodetic measurements is that it fills a gap in the time spectrum between geological observations on the one end and seismic observations at the other end. Geological model predictions can be tested with geodetic measurements and missing elements of the geological models can be filled in, thereby expanding the usefulness of the model concepts for extrapolation to present-day or future tectonic settings as in subduction zones or in sea-level change.

The very nature of many of the deformation phenomena requires observational records that extend over many years. In consequence, the new measurement procedures based on the space-age technologies have not yet made a major impact. Yet new signals are already rising beyond the noise levels and the promises that proponents of the new measurement methods have been making for two decades are now being delivered. It would be hazardous to predict where the new results will lead: new responses to known driving forces will be discovered and new mechanisms will be postulated as developments occur in other areas of the Earth sciences; in seismic tomography or in core dynamics, for example. One reason why this prediction is hazardous is that the levels of observation are now such that they are much contaminated by environmental factors and what is required in order to exploit the new results fully is a parallel program of measuring regional and global atmospheric-oceanic-hydrologic parameters; winds, wind-stress, atmospheric pressure, sea-level and ocean circulation, ground water storage and snow and ice coverage. Much of this would be tedious if it were not for the fact that such data compilations will also advance these environmental sciences, but the rewards are potentially great. The exciting work is only beginning.

References

Cheng MK, Eanes RJ, Shum CK, Schutz BE, Tapley BD (1989) Temporal variations in low degree zonal harmonics from starlette orbit analysis. Geophys. Res. Lett. 16(5): 393-396.

Dziewonski AM (1984) Mapping the lower mantle: determination of lateral heterogeneity in P velocity up to degree and order 6. J. Geophys. Res. 89: 5929-52.

Haxby WG, Karner GD, La Brecque JL, Weissel JK (1983) Digital images of combined oceanic and continental data sets and their use in tectonic studies. EOS, Trans. Am. Geophys. Un 64: 995-1004.

Lambeck K (1980) The Earth's Variable Rotation. Cambridge University Press

Lambeck K (1988) Geophysical Geodesy. Oxford University Press.

Minster JB, Jordan TH (1978) Present-day plate motions. J. Geophys. Res. 83: 5331-54.

NOAA (1973) Reports on geodetic measurements of crustal movement 1906-71. US Dept. Commerce, Nat. Ocean. Atmospher. Admin., Washington DC.

Richards MA, Hager BH (1984) Geoid anomalies in a dynamic Earth. J. Geophys. Res. 89: 5987-6002.

Smith DE, Kolenkiewicz T, Dunn PJ, et al.(1989) The determination of present-day tectonic motions from Laser Ranging to LAGEOS. Paper presented at R. Mather Symposium on Four-Dimensional Geodesy, Univ. NSW, Australia.

Stolz A, Vincent MA, Bender PL, Eanes RJ, Watkins MM, Tapley BD (1989) Rate of change of the Quincy-Monument Peak baseline from a translocation analysis of LAGEOS laser range data Geophys. Res. Lett 16(6): 539-542.

Tsuboi C (1933) Investigation of the deformation of the crust found by precise geodetic means. Jap. J. Astron. Geophys. 10: 93-248.

Wahr JM (1981) Body tides on an elliptical, rotating, elastic and oceanless Earth. Geophys. J. 64: 677-703.

Walcott RI (1984a) The kinematics of the plate boundary zone through New Zealand: A comparison of short- and long-term deformations. Geophys. J. 79: 613-33.

Woodhouse JH, Dziewonski AM (1984) Mapping the upper mantle: three dimensional modeling of Earth structure by inversion of seismic waveforms. J. Geophys. Res. 89: 5953-86.

Absolute Gravimetry as an Operational Tool for Geodynamics Research

W. Torge
University of Hannover
F.R.G.

ABSTRACT:

Relative gravimetric techniques have been used for nearly 30 years for measuring non-tidal gravity variations with time, and thus have contributed to geodynamics research by monitoring vertical crustal movements and internal mass shifts. With today's accuracy of about $\pm$ $0.05\mu ms^{-2}$ (or $5\mu Gal$), significant results have been obtained in numerous control nets of local extension, especially in connection with seismic and volcanic events. Nevertheless, the main drawbacks of relative gravimetry, which are deficiencies in absolute datum and calibration, set a limit for its application, especially with respect to large-scale networks and long-term investigations.

These problems can now be successfully attacked by absolute gravimetry, with transportable gravimeters available since about 20 years. While the absolute technique during the first two centuries of gravimetry's history was based on the pendulum method, the free-fall method can now be employed taking advantage of laser-interferometry, electronic timing, vacuum and shock absorbing techniques, and on-line computer-control. The accuracy inherent in advanced instruments is about $\pm 0.05 \ \mu ms^{-2}$. In field work, generally an accuracy of $\pm 0.1 \ \mu ms^{-2}$ may be expected, strongly depending on local environmental conditions.

The efficiency of transportable absolute gravity measuring systems is demonstrated by the experiences collected with the JULAG-3 gravimeter. This Faller-type instrument has been operated by the Institut für Erdmessung since more than three years. After laboratory and fields tests and some modifications, it has been employed for approximately 100 absolute gravity determinations, on about 60 different stations in central and northern Europe, Greenland, and South America. While some of these stations contribute to the global absolute gravity network, most of them serve for regional and local gravimetric control, especially for geodynamic investigations at tectonic plate boundaries (Iceland, Venezuelan and Central Andes), and at intraplate seismic (south western Germany) or uplift/subsidence (Denmark, northern Germany) areas. On an average station a precision of a few $0.01 \ \mu ms^{-2}$ can be achieved from 1500 individual drops, within two days. The accuracy as estimated by repeated measurements and comparison with the result of other instruments, is at least $\pm 0.1 \ \mu ms^{-2}$. Further improvements aim to increase the accuracy by a factor of two.

1. INTRODUCTION

Repeated gravity measurements are one tool for detecting mass shifts in the earth, with special power to monitor vertical movements of the earth's surface and in the crust (Torge 1986). With respect to vertical crustal movements, a height change of $+1$ cm generally corresponds to a gravity change of $- 0.02 \ldots - 0.03\ \mu ms^{-2}$ ($1\ \mu ms^{-2} = 0.1$ mGal), depending on the local behaviour of mass transfer. In geodynamics research, gravity changes related to recent tectonic processes are of interest. At global and regional scale, long-time variations at the order of 0.01 to 0.1 $\mu ms^{-2}/a$ have to be expected, while more local changes related to seismo-tectonics and volcanic activities elapse over time spans of days to some years, with variations reaching 1 μms^{-2} and more. In order to significantly detect these changes, gravity values have to be determined with an accuracy of $\pm$ 0.01 to 0.1 μms^{-2}. Control point stability with time as well as gravity reduction for non-tectonics effects should also beat this accuracy level.

For nearly 30 years, relative gravimetry has been successfully employed in local investigations of geodynamic processes, concentrating on test areas at tectonic plate boundaries. For monumented control points occupied several times with high-precision spring gravimeters especially of type LaCoste and Romberg (LCR), relative accuracies of some $\pm 0.01\ \mu ms^{-2}$ may be reached. One example of such a local control is the northeastern Iceland gravity network (Torge and Kanngieser 1983). Following the early work of A. Schleusener, the Institut für Erdmessung (IfE), Universität Hannover, has in regular intervals between 1965 and 1987, observed gravity and height variations at that part of an accreting plate boundary. By continuous improvement of instrumentation as well as observation, reduction and evaluation methods, an average station accuracy of $\pm 0.06\ \mu ms^{-2}$ has been finally achieved. But, relative gravimetric techniques with spring gravimeters exhibit several drawbacks resulting from the use of an elastic counterforce in order to measure gravity :

- as absolute gravity level cannot be determined, a datum deficiency enters into gravity networks. This fact poses severe problems when observation results of different epochs are compared, and forces us to introduce assumptions about the time stability at certain network parts,

- as gravity differences are derived from length changes of the elastic spring, a calibration problem exists. Because of the complex transfer from the gravimeter's reading device to the sensor, the calibration function has a complicated structure which makes it difficult to reach the 0.1 μms^{-2} accuracy level for large-scale networks,

- as the elastic force is subject to changes with time, repeated and overlapping observations have to be performed in order to obtain network homogeneity and to control error propagation. This fact may significantly increase the expenses of the network survey.

With the introduction of transportable absolute gravimeters about 20 years ago, those problems could be attacked successfully. The last generation of these devices has now reached a certain operational state, delivering the $\pm 0.1\ \mu ms^{-2}$ accuracy. We shortly summarize in the following, the principles and the status of absolute gravimetry. Using the example of JILAG-3, a Faller-type absolute gravimeter operated by IfE since 1986, we then demonstrate in more detail the potential of this technique at establishing high-precision gravity control systems of global, regional and local scale. Concentrating on the accuracies achieved, we describe the experiences obtained in several projects under different environmental conditions, and point to open problems.

2. DEVELOPMENT AND PRESENT STATE OF ABSOLUTE GRAVI-METRY

With the law of free fall and the pendulum law, Galileo Galilei about 1600 established two basic relationships for gravimetry. Until the end of the 19th Century, only pendulum instruments designed as wire, or later also as reversible pendulums, were used for gravity measurements. With the constructions of Repsold and the Brunner Brothers, even transportable reversible pendulums were employed in the second half of the 19th Century. After the introduction of v. Sterneck's relative pendulum apparatus in 1887, and with the development of elastic spring gravimeters since about 1930, relative techniques have dominated gravimetry until recently [1].

A remarkable change happened in the 1950's, when the free-fall method could for the first time be employed due to the progress in short-time measurement technique. About 1970, laser interferometry, vacuum technique and advanced vibration isolating devices allowed gravimeters to reach the 0.1 μms^{-2} precision level. The erá of transportable free-fall gravimeters was opened by the development of J.E. Faller, which successfully operated at the establishment of the International Gravity Standardization Net 1971 (Hammond and Faller 1971). Meanwhile, about 12 transportable systems have been developed, based either on the free-fall or the symmetric rise-and-fall principle, and brought into operation.

At all these developments simultaneous distance and time measurements are performed using the Michelson interferometer principle with laser light (Fig. 1). One of the corner cube reflectors forming the interferometer can be moved in the vertical, thus representing the gravity sensor. The other reflector is fixed and serves as a reference. The light from the stabilized He-Ne-gaslaser ($\lambda = 632.8$ nm) is split into a reference beam and a measurement beam. At the corner cube reflectors, the beams are reflected parallel to the incoming light and again passed through the beam splitter, where they are superimposed. Shifting the falling body by $\lambda/2$ leads to interference fringes at the photodetector. This optical signal is converted into an electic signal, amplified and triggered, and the number of fringe impulses is registered with an electronic counter. With n impulse counts, the distance covered is $n * \lambda/2$. Timing is controlled by an atomic frequency standard. Atmospheric pressure effects are largely reduced by performing the experiment in vacuum, while microseismic ground movements influencing the reference corner cube are diminished by one to two orders of magnitude through long-period compensation devices. At more advanced instruments, 100 to 1000 single time measurements are made at one experiment, at preset fringe counts, and an on-line evaluation is done with a micro-computer,which is also used for the control of the measurement

---

[1] A detailed synopsis about gravity measurement systems is given by Torge (1989).

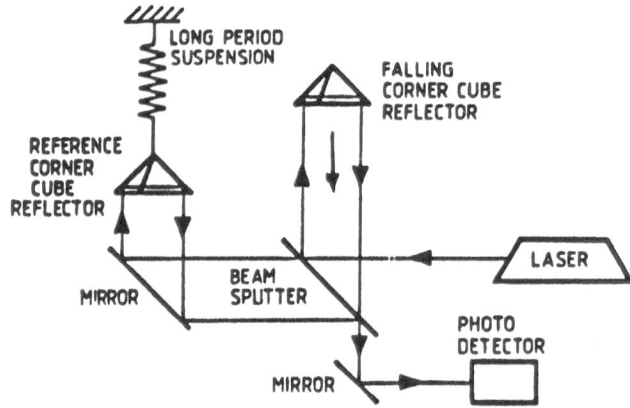

Fig. 1. Michelson interferometer system at free-fall method

process. In order to further reduce residual microseismics, and filter out short-term gravity changes, a large number (100 to 1000 or more) of experiments are made in one station, taking a total occupation time of one to several days per station.

Regarding the accuracy which can be reached with present-day absolute gravimeters, there exist several estimates. They stem from the analysis of individual error components, as well as from repeated observations on the same station, and from comparisons with the results of other instruments. An a priori error budget first includes random errors in the resolution of the interference-fringe signal and timing errors in associating this signal with time impulses, as well as most part of the residual microseismics. This random error part can be estimated from the drop-to-drop scatter. Systematic errors are either of instrumental character (standards used, phase shift at electronic timing, deviation of the optical path from ideal conditions, residual air pressure, temperature gradient and non-gravitional forces, floor recoil effects on instable floor), or they represent uncertainties of gravity field reductions. These reductions model certain gravity changes with time (earth tides, polar motion, changes of air pressure), and they transfer the observed gravity value to a ground mark. Tab. 1 summarises the error budget for advanced transportable free-fall gravimeters, valid under average station conditions, as compiled from the literature (Faller and Marson 1988, Arnautov et al. 1989). In the listing, enviromental effects have been included, which enter into comparisons if the same station is occupied at another time. As groundwater and soil moisture changes generally cannot be corrected, the environmental error part may be increased by amounts between 0 (bed-rock station) and 0.05 μms^{-2}, or more.

Tab. 1. Average error budget of advanced transportable free-fall gravimeters, after Faller and Marson (1988) and others.

error source	contribution $\mu m s^{-2}$
instrumental	
random error part	± 0.030
laser wavelength	± 0.010
frequency standard	± 0.005
electronic phase shift	± 0.015
optical path deviations	± 0.015
differential pressure	± 0.020
differential temperature	± 0.010
magnetic field, electrostatics	± 0.010
floor recoil	± 0.015
r.m.s. subtotal	± 0.048
environmental	
earth tide reduction	± 0.010
polar motion reduction	± 0.005
air pressure reduction	± 0.005
reduction to ground	± 0.020
r.m.s. total	± 0.053

From repeated measurements performed with the same instrument in the same station, different values for the repeatability (r.m.s. scatter about mean value) have been derived. At carefully selected stations in the U.S.A., reoccupied after an interval of a few days to one year, values of ±0.01 to 0.04 $\mu m s^{-2}$ have been found for the JILAG-4 gravimeter (Peter et al. 1989). JILAG stands for the most recent version of a free-fall gravimeter, constructed at the Joint Institute for Laboratory Astrophysics, University of Colorado and National Institute of Standards and Technology, Boulder, by Prof. J.E. Faller and coworkers. This gravimeter type being built in a series of six instruments between 1983 and 1986, and operated by different institutions worldwide, stands for one of the most advanced free-fall devices (Faller et al. 1983). For JILAG-3, operated by IfE Hannover, a r.m.s. repeatability of ± 0.08 $\mu m s^{-2}$ has been found over 3 years for the station Hannover, located in the pleistocene sedimentary basin of northern Germany (see chapter 3). If different instruments are compared in the same station, the r.m.s. scatter about the mean value represents not only the instruments precision and certain instrumental and environmental effects which change with time, but also systematic differences between the different gravimeters, and observation and evaluation procedures. For evaluating the accuracy of one absolute determination, we may use the results of 19 different measurements, made with 8 different gravimeters between 1976 and 1986 in the Bureau International des Poids et Mésures (BIPM), Sèvres, mainly within two international comparison campaigns (Boulanger et al. 1986). Taking all results with equal weight and assuming a random error behaviour for the scatter about the mean value, we get a standard deviation of $\pm$ 0.11 $\mu m s^{-2}$ for an individual result (Torge 1987).

We consequently may conclude, that present-day absolute gravimeters may be able to reach a $\pm$ 0.05 $\mu m s^{-2}$ accuracy, but that an actual accuracy of about $\pm$ 0.1 $\mu m s^{-2}$ is achieved, although this figure will change depending on station conditions and environmental effects.

3. EXPERIENCES WITH THE JILAG-3 ABSOLUTE GRAVIMETER

Since January 1986, the Institut für Erdmessung (IfE), Universität Hannover, has operated the JILAG-3, one of the instruments constructed by Prof. J.E. Faller and coworkers at JILA (see chapter 2). After laboratory and field experiments, hard- and software improvements, as well as comparative measurements at absolute and relative gravity stations, the gravimeter has been employed for absolute gravity control in fundamental and geodynamic networks in central and northern Europe (Torge et al. 1987, 1988), and in South America (Gemael et al. 1989). In connection with these projects, also four stations of the International Absolute Gravity Basestation Network (IAGBN) proposed by the International Association of Geodesy (Boedecker and Fritzer 1986) have been established. Altogether, approximately 100 absolute gravity determinations at about 60 different stations have been carried out. Tab. 2 gives an overview of the projects performed with JILAG-3 by IfE, between 1986 and 1989, with number of observed stations. Fig. 2 and Fig. 3 show the location of the absolute gravity stations established in northern Europe and Greenland, and in South America. The JILAG-3 gravimeter is operated by two observers, and under normal transportation and station conditions, it delivers reliable results within two days, including assembling and disassembling. At each drop, 150 to 200 time-distance measurements are fitted to a parabola, yielding one gravity value. The standard deviation for one drop strongly depends on local station conditions and may vary between 0.2 and 2 $\mu m s^{-2}$. With 500 to 2000 drops, an average station precision of a few 0.01 $\mu m s^{-2}$ can be achieved. Fig. 4 shows the drop to drop scatter and histograms for a station with low (Brasilia) and high (Vicosa) local noise. With respect to the systematic errors described in chapter 2, the laser stability is about $2*10^{-9}/a$, and regularly controlled by a iodine-stabilized laser. The rubidium frequency standard has a long-term stability of better than $10^{-10}/a$. The instrument is operated in a vacuum of 10^{-4} Pa. Corrections for earth tides (constant amplitude factor 1.164, phase shift zero), polar motion, and air pressure (-3 nms^{-2}/hPa) are applied.

Tab. 2. Absolute gravimetry projects performed with JILAG-3, IfE Hannover.

1986/87	F.R.G. base net	11	instr. test and network control
1986	Danish land uplift line	3	geodynamics
1986	BIPM Sèvres	3	comparison
1987	Calibration System Hannover	7	instr. test and calibr. line control
1987	Obs. Royal Brussels	1	drift control of supercond. gravim.
1987	northern Iceland	6	geodynamics (rift zone)
1988	ZIPE Potsdam	1	comparison
1988	Faeroer, Iceland, Greenland	5	absol. control, IAGBN
1988	South-west Germany	8	geodynamics (seismic area)
1988	Venezuela	6	abs. control, geodynamics (Venez. Andes), IAGBN
1988	West-Berlin	1	abs. control
1989	Brasil, Uruguay, Argentine	12	abs. control, geodynamics (central Andes), IAGBN

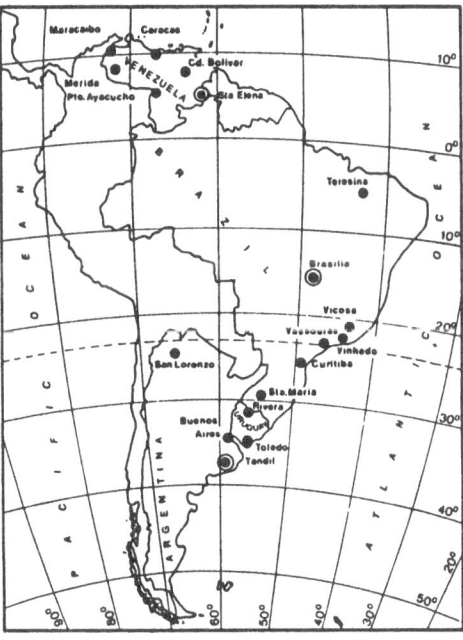

Fig. 2 (left). Absolute gravity stations established with JILAG-3 in northern Europe and Greenland, IAGBN station Godthab.
Fig. 3 (right). Absolute gravity stations established with JILAG-3 in South America, IAGBN stations Sta. Elena, Brasilia, Tandil.

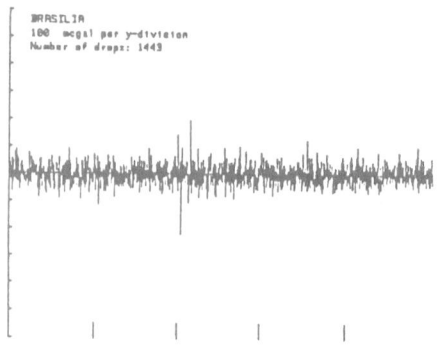
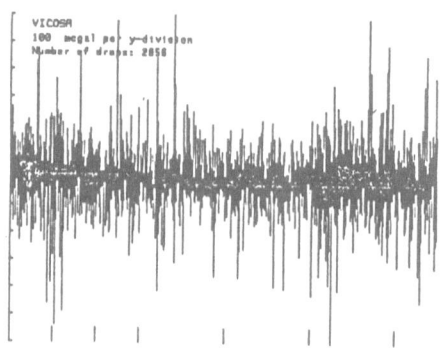

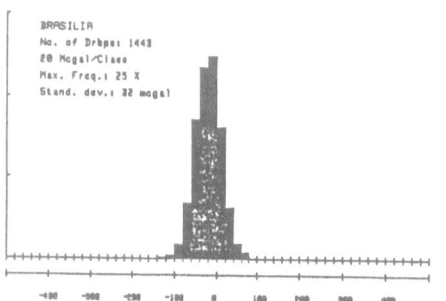
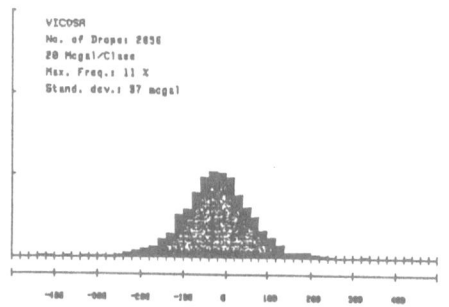

Fig. 4. Drop to drop scatter and histograms for JILAG-3, at a station of low noise (left): Brasilia, 1443 drops, standard deviation $\pm$ 0.32 $\mu ms^{-2}/$ drop, and at a station of high noise (right): Vicosa, 2856 drops, standard deviation $\pm$ 0.97 μms^{-2}.

Instrumental corrections include the time delay caused by the finite light velocity, and a temperature dependent laser wavelength term. The readjustment of the instrument's interferometer base in 1988 resulted in a correction of $+$ 0.22 μms^{-2}, to be applied to all results obtained before February 1988 (Torge et al. 1989). The gravity difference between the reference height (about 80 cm) to a ground mark is measured ten times with each of two LCR gravimeters equipped with electrostatic feedback systems developed at IfE.

The accuracy achieved with JILAG-3 has been estimated from the scatter of repeated measurements, and by comparisons with other instruments.

In the station Hannover 101, which is not the most favourable one with respect to local disturbance, the repeatability may been estimated from 16 determinations performed between 1986 and 1989, resulting in a r.m.s. scatter of $\pm$ 0.08 μms^{-2} (Torge et al. 1989), see Fig. 5.

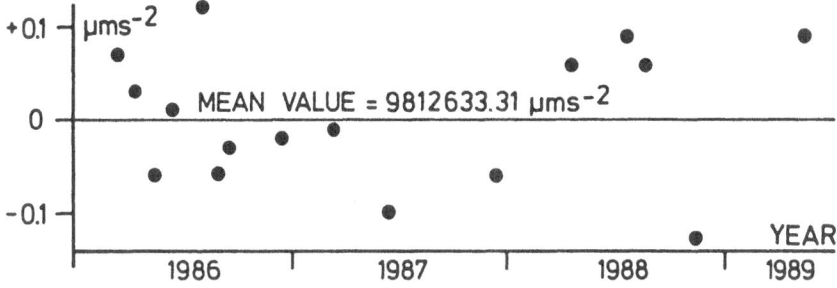

Fig. 5. Repeatability of JILAG-3 absolute gravity determinations, station Hannover 101, 1986 – 1989.

A large effort has been undertaken to investigate the accuracy of JILAG-3 results, by comparing them with the gravity values obtained with other instruments.

A comparison with a high-precision relative gravimetry network was possible with the Gravity Base Net 1976 (DSGN76) of the Federal Republic of Germany (Sigl et al. 1981). This network has been observed repeatedly in 1977 with 4 LCR-gravimeters, giving 656 gravity differences between the 21 network stations. Based on four absolute determinations performed by the Instituto di Metrologia "G. Colonnetti", Torino (Cannizzo et al. 1978), the standard deviation of the adjusted gravity values are ± 0.06 ... 0.11 μms^{-2}. Fig. 6 shows the northern part of DSGN76, the 1. order densification net of lower Saxony, and the gravimeter calibration line Cuxhaven-Hannover-Harz (Kanngieser et al. 1983),together with JILAG-3 absolute gravity determinations.

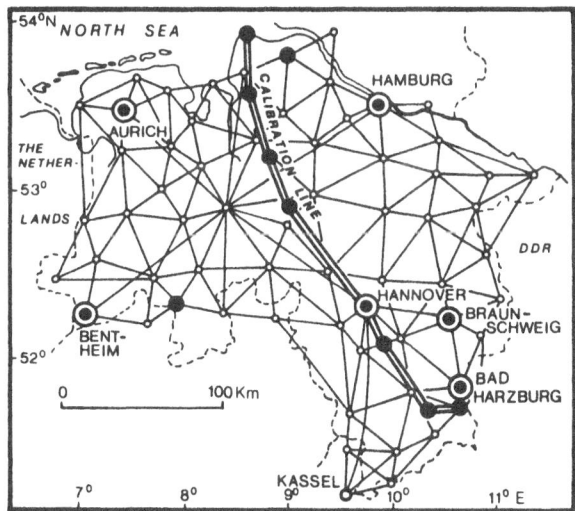

Fig. 6. Northern part of DSGN76 (with station names), 1. order gravity net of lower Saxony, and calibration line Cuxhaven-Hannover-Harz, with JILAG-3 absolute gravity stations (black circles).

The comparison between the JILAG-3 results, corrected by $+ 0.22 \ \mu m s^{-2}$ (see above), and the DSGN76-values, is given in Tab. 3.

Tab. 3. Comparison of JILAG-3 results with DSGN76, JILAG-3 values centered to DSGN76, (2) = two independent determinations.

Station	Gravity ($\mu m s^{-2}$)		
	JILAG-3	DSGN76	$g_{JILAG-3} - g_{DSGN76}$
München	9807231.33	... 1.29	+ 0.04
Wiesbaden	9810368.67	... 8.64	+ 0.03
Bad Harzburg	9811655.17	... 5.20	− 0.03
Braunschweig(2)	9812529.44	... 9.43	+ 0.01
Hannover	9812624.03	... 4.04	− 0.01
Bentheim	9812706.38	... 6.40	− 0.02
Aurich(2)	9813567.48	... 7.67	− 0.19
Hamburg(2)	9813636.73	... 6.79	− 0.06
			− 0.03

We note, that there is an excellent absolute and relative agreement between JILAG-3 and DSGN76, expressed by a r.m.s. discrepancy of $\pm \ 0.07 \ \mu m s^{-2}$. As there is practically no constant bias, the absolute datum of DSGN76 has been perfectly fixed by the IMGC-values for München, Wiesbaden, Braunschweig, and Hamburg, although the direct comparison of the absolute values reveals larger discrepancies (see Tab. 5). This seems to be due to the strong relative ties in DSGN76, and a reasonable weighting between absolute and relative data at the DSGN76 adjustment. From the discrepancies we also find, that the accuracy of gravity differences observed with JILAG-3 is rather high, it may even be reduced to a r.m.s. discrepancy of $\pm \ 0.03 \ \mu m s^{-2}$ if we take the station Aurich out. This fact hints to systematic instrumental effects, which cancel in the difference between two stations. Another comparison with high-precision relative gravimetry, calibrated in the Hannover Calibration System, was possible with the gravity network established by IfE in Iceland (Torge and Kanngieser 1983). The absolute level of this network has been derived from numerous gravity connections performed with several LCR gravimeters in the time span 1965 to 1985, between the Hannover Calibration System and Reykjavik (56 gravity differences), and Reykjavik and Akureyri (65 differences). After applying a shift of $+ \ 0.22 \ \mu m s^{-2}$ to the calibration system, in order to fit it to the JILAG-3 results of 7 stations in Central Europe, the results given in Tab. 4 have been obtained (2 absolute stations observed 1987 in northern Iceland have been omitted due to instrumental problems). The gravity transfer accuracy to the Iceland network had been estimated before to $\pm \ 0.1 \ ... \ 0.2 \ \mu m s^{-2}$, while the relative accuracy within the northern Iceland network should be better than $\pm \ 0.1 \ \mu m s^{-2}$. Absolute values have been determined 1987 on 6 stations in northern Iceland and 1988 in Reykjavik.

Tab. 4. Comparison of JILAG-3 results with the gravity network northern Iceland.

Station	Gravity (μms^{-2})		
	JILAG-3	Iceland net	$g_{JILAG-3} - g_{Iceland}$
Reykjavik 1062	9822647.82	... 7.84	− 0.02
Akureyri 60932	(9823334.12) transfer from Reykjavik	... 4.10	+ 0.02
Gardur 1041	9823825.67	... 5.58	+ 0.09
Husavik 1021	9823703.99	... 3.77	+ 0.22
Krafla 1051	9822528.59	... 8.61	− 0.02
Laugar 1011	9823399.59	... 9.67	− 0.08

We note, that even under the difficult environmental conditions in the northern Iceland network, an ± 0.1 μms^{-2} accuracy could be maintained.

The most stringent quality control is provided by comparisons with other absolute gravimeter results. This was first possible by a measurement performed 1986 at the international reference station BIPM / Sèvres (Boulanger et al. 1986, see chapter 2).

From 18 absolute determinations, made with 7 other gravimeters between 1976 and 1985, a *formal* error calculation assuming equal weights, gives a standard deviation of ± 0.12 μms^{-2} for the individual determination, and ± 0.03 μms^{-2} for the mean value (Torge 1987). Observations done in 1986 and 1987 on stations in Central Europe occupied in 1976/77 with the IMGC apparatus, with an estimated ±0.1 μms^{-2} accuracy (Cannizzo et al. 1978) gave another opportunity for evaluating the JILAG-3 accuracy. As the IMGC-vertical gradients are supposed to contain larger errors, IMGC values have been reduced to ground floor with gradients measured by IfE (Torge et al. 1987,1988). Finally a JILAG-3 measurement was carried out at Zentralinstitut für Physik der Erde (ZIPE), Potsdam, in 1988. In this station of the classical absolute determination around 1900, taken as base for the former Potsdam Gravity System, 7 absolute determinations have been performed with the GABL free-fall gravimeter of USSR, between 1976 and 1986 (Arnautov et al. 1989). The accuracy of the GABL results for the floor station has been estimated to ± 0.04 ... 0.17 μms^{-2}. Although a slight gravity variation with time is indicated in the results, we have for the comparison taken the equal weighted mean value for 1976 to 1986, with a calculated standard deviation of ± 0.08 μms^{-2}. The results of these comparisons (JILAG-3 values corrected by + 0.22 μms^{-2}) are given in Tab. 5.

Tab. 5. Comparison of JILAG-3 results with absolute gravity values obtained with other gravimeters.

Station	Gravity ($\mu m s^{-2}$)			Remarks
	JILAG-3	Ref.Grav.	$g_{JILAG-3} - g_{Ref.Grav.}$	
BIPM Sèvres A	9809259.97	... 9.93	+ 0.04	Ref: mean value 1976-1985 from 7 instruments
München	9807231.33	... 1.28	+ 0.05	IMGC-results 1976/77 reduced with IfE gradients
Sèvres A3	9809259.35	... 9.17	+ 0.18	
Wiesbaden	9810368.67	... 8.67	0.	
Brussels A	9811172.87	... 2.66	+ 0.21	
Braunschweig	9812529.44	... 9.23	+ 0.21	
Hamburg	9813636.73	... 6.92	− 0.19	
Copenhagen	9814956.13	... 5.82	+ 0.31	
ZIPE Potsdam S14 ·	9812616.73	... 6.78	− 0.05	Ref: mean value 1976-1988 of GABL-results

From the absolute comparisons of Tab. 5, we find excellent agreement with other results, if the reference value has been derived as a long-term average (over 10 years or more), as in the stations Sèvres and Potsdam. The comparison with the IMGC values, obtained about 10 years before the JILAG-3 data, indicates a systematic bias of $+ 0.11 \ \mu m s^{-2}$, and gives a r.m.s. discrepancy of ± 0.19 resp. (bias reduced) ± 0.17 $\mu m s^{-2}$. This points to some systematic effects of instrumental or local environment source, which increase the a priori estimates of $\pm 0.1 \ \mu m s^{-2}$, assumed for both data sets. A slightly better result has been found at the comparison of NGS-JILAG-4 results (1987/88) with other absolute values on four stations in the U.S.A. (Peter et al. 1989). If four determinations (IMGC, AFGL) with discrepancies of $0.5 \dots 0.7 \ \mu m s^{-2}$ are excluded, the r.m.s. discrepancy derived from comparing with six JILA and one AFGL value (1980/ 1982) is $\pm 0.14 \ \mu m s^{-2}$ before and $\pm 0.11 \ \mu m s^{-2}$ after bias-reduction, with a bias (JILAG-4 − Ref.Grav.) of $+ 0.09 \ \mu m s^{-2}$.

4. CONCLUSION

From the experiences collected during the last three years with the JILAG-3, an advanced transportable absolute gravimeter, and from similar results obtained with other instruments we may state that

- the attainable accuracy of a gravity determination as defined by instrumental properties and reduction uncertainties, is about $\pm$ 0.05 $\mu m s^{-2}$, which is compatible with the accuracy of relative gravity networks at local scale,

- the achieved accuracy is about $\pm$ 0.1 $\mu m s^{-2}$ on the average, strongly depending on local conditions, with eventual contributions also from systematic instrumental errors,

- the instruments have reached a certain operational state, and can be used even under difficult transportation and environment conditions, with a reliable gravity determination performed within a few days.

Although further improvements in hard- and software, as well as in reduction procedures are necessary in order to reach the 0.05 $\mu m s^{-2}$ accuracy inherent in the method, the available instruments can already be successfully employed for geodynamics research in

- establishing global and large-scale regional absolute gravity nets, particularly in connection with geodetic space techniques control,

- supporting regional and local relative gravimetry networks, by providing the absolute datum and linear scale, and improving network accuracy and reliability through control of error propagation,

- improving gravimeter calibration lines with respect to long-wave calibration parameters,

- strengthening and extending height control systems established through geometric terrestrial or space methods, for the investigation of recent crustal movements.

5. REFERENCES

Arnautov GP, Boulanger YuD, Šeglov SN, Elstner Cl (1989) Absolute gravity measurements at Potsdam – Results of the gravimeter GABL. Gerlands Beiträge zur Geophysik (in press).

Boedecker G, Fritzer Th (1986) International Absolute Gravity Basestation Network. Veröff. Bayer. Komm. für die Internat. Erdmessung der Bayer. Akad. d. Wissensch., Astron.-Geod. Arb., Heft Nr. 47, München.

Boulanger YuD, Faller JE, Groten E et al. (1986) Results of the second international comparison of absolute gravimeters in Sèvres 1985. Bur. Grav. Int., Bull. d'Inf. No. 59: 89-103.

Cannizzo L, Cerutti G, Marson I (1978) Absolute gravity measurements in Europe. Il Nuovo Cimento 1C: 39-85.

Faller JE, Marson I (1988) Ballistic methods of measuring g – the direct free-fall and symmetrical rise-and fall methods compared. Metrologia 25: 49-55.

Faller JE, Guo YG, Gschwind J, Niebauer TM, Rinker RC, Xue J (1983) The JILA portable absolute gravity apparatus. Bur. Grav. Int., Bull. d'Inf. No. 53: 87-97.

Gemael C, Leite OHS, Rosier FA, Torge W, Röder RH, Schnüll M (1989) Large-scale absolute gravity control in Brasil. Pres. to IAG General Meeting Edinburgh, Scotland, 03.-12. August 1989. Proceedings Springer Int. (in press).

Hammond JA, Faller JE (1971) Results of absolute gravity determinations at a number of different sites. J. Geophys. Res. 76: 7850-7854.

Kanngieser E, Kummer K, Torge W, Wenzel HG (1983) Das Gravimeter-Eichsystem Hannover. Wiss. Arb. Fachr. Verm.wesen, Univ. Hannover, Nr. 120.

Peter G, Moose RE, Wessels CW, Faller JE, Niebauer TM (1989) High-precision absolute gravity observations in the United States. J. Geophys. Res. 94: 5659-5674.

Sigl R, Torge W, Beetz H, Stuber R (1981) Das Schweregrundnetz 1976 der Bundesrepublik Deutschland (DSGN76), Teil I. D. Geod. Komm., Reihe B, Nr. 254, München.

Torge W (1986) Gravimetry for monitoring vertical crustal movements : potential and problems. Tectonophysics 130: 385-393.

Torge W (1987) Absolute Schweremessung mit transportablen Gravimetern – ein Umbruch in der Gravimetrie. Z. f. Verm.wesen 112: 224-234.

Torge W (1989) Gravimetry. W. de Gruyter, Berlin New York.

Torge W, Kanngieser E (1983) Gravity and height variations connected with the recent rifting process in northern Iceland. J. Geophys. 53: 24-33.

Torge W, Röder RH, Schnüll M, Wenzel HG, Faller JE (1987) First results with the transportable absolute gravity meter JILAG-3. Bull. Gèod. 61: 161-176.

Torge W, Röder RH, Schnüll M, Wenzel HG, Faller JE (1988) Laboratory and field tests of JILAG-3 absolute gravimeter. Bur. Grav. Int., Bull. d'Inf. No. 62: 36-40.

Torge W, Röder RH, Schnüll M (1989) Correction of a systematic error of gravity measurements with JILAG-3. Bur. Grav., Int. Bull. d'Inf. (in press).

Satellite Gravity Gradiometry: A Future Technique for Global High Resolution Gravity Field Recovery

K.H. Ilk
Technical University Munich
F.R.G.

ABSTRACT:

One future possibility for obtaining improved high resolution gravity field information, besides the competitive technique of satellite-to-satellite tracking, is satellite gravity gradiometry. In this paper the current state of the art of gravity field determination is sketched as well as the need for an improvement of our knowledge of the gravity field. Then the structure of the tidal field is discussed and the basic mechanical principles of satellite gravity gradiometry are outlined. The size of the gravity signal which gravity gradiometers have to detect is illustrated and the relationship between instrumentation accuracy versus gravity field parameter recovery accuracy is discussed. The results of a detailed simulation study, restricted to a regional recovery example, is presented.

1. A SKETCH OF THE CURRENT STATE OF THE ART

The constant improvement in quality, quantity and distribution of satellite observations during the past three decades together with a high
quality terrestrial gravity material covering many parts of the world
has provided the basis for a global modelling of the gravity field of
the Earth with reasonable accuracy.

Satellite tracking methods contribute to these gravity models mainly in
the low-frequency range with a resolution of up to some thousands of
kilometers. Satellite altimetry and improved terrestrial gravity
data supply the mean and high frequency part of the gravity field
spectrum with a resolution of a couple of hundreds of kilometers. The
present global gravity field models, spherical harmonics expansions
complete up to degree 36, e.g. GRIM 3-L1 (Reigber et al. (1985)) or
GEM 10B (Lerch et al.(1981)) have a resolution of 550 km at the equator and an accuracy of 2 to 3 m in terms of geoid heights and of 5 mgal
in terms of 5° x 5° gravity anomalies. Besides these, combined solutions based on satellite tracking and terrestrial gravity data, pure
satellite solutions have been derived for special tasks, e.g. GEM-T1
(Marsh et al.(1987)) or high-resolution gravity models, spherical harmonics expansions complete up to degree and order 360, where the low
frequency part is derived again from satellite tracking data and the
short wavelength features are provided by satellite altimetry and by
surface gravity information. The latter gravity models describe the
gravity field with an accuracy of 20 mgal in terms of 1^O x 1^O gravity
anomalies on the continents and 8 mgal in the ocean areas; the spatial
resolution is 50 to 100 km at the equator.

Despite this progress in gravity field modelling the use for scientific
and practical applications in geosciences is limited because of still
limited accuracy and insufficient spatial resolution of the recovered
gravity field. A distinct improvement using present day observation
sets and types of observations does not seem possible - only marginal
refinements applying improved processing procedures or the inclusion
of additional satellite tracking data are possible. The reasons are
inherent in the terrestrial gravity material on the one hand and in the
satellite data on the other hand. The terrestrial data suffer from insufficient coverage due to operational and political reasons and from
still limited accuracy in many parts of the world. Satellite tracking
data are contaminated by force function uncertainties which cannot be
modelled with sufficient accuracy - mainly atmospheric effects - and do
not cover the satellite arcs with sufficient density. Moreover the

satellite altitudes are too high because of the above mentioned reasons and, therefore, show a too low signal to noise ratio.

To find meaningful answers to important questions related to solid-
earth dynamics, ocean dynamics and physical geodesy it is necessary to
know the gravity field with an improved accuracy and a much higher
resolution. Without going into detailed requirements for the different
disciplines one can state that for most of the objectives global homogeneous gravity information with an accuracy of 2 to 5 mgal in terms of
gravity anomalies and 5 to 10 cm in terms of geoid heights and with an
average spatial resolution of 50 to 100 km would be adequate.

Such an improved knowledge of the gravity field with respect to resolution, accuracy, and even with the possibility of a timelike repeatability must come from entirely new data types. These data types should
cover the complete range of the gravity field spectrum and should be
collected all over the Earth. Only a dedicated satellite gravity mission where the gravity sensor is flown in a polar orbit at an altitude
as low as possible could provide such gravity field information within
a reasonable time period. There are two competitive techniques under
discussion: the satellite-to-satellite tracking (SST) technique, either
in the low-low or in the high-low mode, and the satellite gravity gradiometry (SGG), where all nine elements of the gravity tensor or a par-
tition are measured simultaneously.

From the physical point of view all these methods are based on the same
principle: the relative motion of two or more test particles are measured (e.g. relative velocity or relative acceleration) and analysed for
the unknown gravity field components. Typical test particle separations
of up to one meter are planned to be used for SGG, of up to a couple of
hundreds of kilometers for the SST low-low mode and up to several thou-
send kilometers for the SST high-low mode.

Actually, the idea of satellite gravity gradiometry seems to be the
more attractive technique compared to satellite-to-satellite tracking
because of lower costs and higher benefit. There are three missions
proposed for the 1990's, two SGG missions and one SST mission. The SGG
missions are the Aristoteles Mission, the first project of ESA's Solid
Earth Programme, and NASA's Superconducting Gravity Gradiometer Mission (SGGM). The SST mission is NASA's Geopotential Research Mission
(GRM) a SST low-low mode approach with two co-orbiting drag free satellites in circular polar orbits at an altitude of 160 km and a separation between 150 to 550 km (Keating et al., 1986). The Aristoteles-
Mission is planned to consist basically of a two-axis shielded gradio-

meter, the GRADIO instrument, consisting of four accelerometers each with two highly sensitive axis (radial and cross)(Balmino et al.,1985). These accelerometers have a sensitivity of 10^{-11} ms s^{-2} $Hz^{-1/2}$, and an interaxis orthogonality and interaxis alignment accuracy of 10^{-6} rad. The satellite will operate for half a year at an altitude of about 200 km in a near-polar orbit. The launch is planned for the mid-1990's. The SGGM is currently proposed by NASA for the second half of the next decade. This very ambitious mission consists of a three-axis superconducting gravity gradiometer presently under development at the University of Maryland. The sensitivity of this apparatus is expected to be 10^{-4} E $Hz^{-1/2}$ (1E.U. = 10^{-9} sec^{-2}) and limited only by thermal and Superconducting QUantum Interference Device (SQUID) ampli-fier noise (Paik et al., 1988). The instrument is intended to be flown in the inner component of a drag free satellite in a circular polar orbit at an altitude of 160 km to 200 km. A gradiometer with a 0.2 m separation and a sensitivity as mentioned above must have an attitude control of $3.2 \cdot 10^{-7}$ rad s^{-1}, one of the most demanding tasks of the mission.

In the following we will not go into further technical details related to these missions planned so far, or to missions discussed in the past. For further information concerning measurement principles of various types of gravity gradiometers we refer to (Spaceborne Gravity Gradio-meters, 1983). We will rather discuss the mechanical principles on which gravity gradiometry is based. Further we will illustrate the size of signals which gravity gradiometers have to detect and the relation of instrumentation accuracy versus recovery accuracy to get an idea of the challenge the engineers are faced with. The simulation examples given in this paper are restricted to regional gravity field mapping - rather due to computation effort reasons than due to the conviction that the regional mapping could really be an alternative to the global mapping. The long-term aim is, of course, the precise global mapping of the gravity field characterised by the accuracy numbers given above. But we believe that a complex global solution can only be derived by a sequential procedure. It seems that the merging of region-al results to a global solution has some advantages over a one-step global solution - but this is just one of the many open questions related to these future gravity field mapping methods.

2. THE STRUCTURE OF THE TIDAL FIELD AND ITS DETECTION

As mentioned above, the mechanical principles of SST and SGG are quite similar. In a certain respect the SST principle can be considered as sort of one-axis gravity gradiometer. To illustrate the principle one has to remember that the gravity field of the Earth acts in a free falling spacecraft in the form of the tidal field which is shown in fig.1

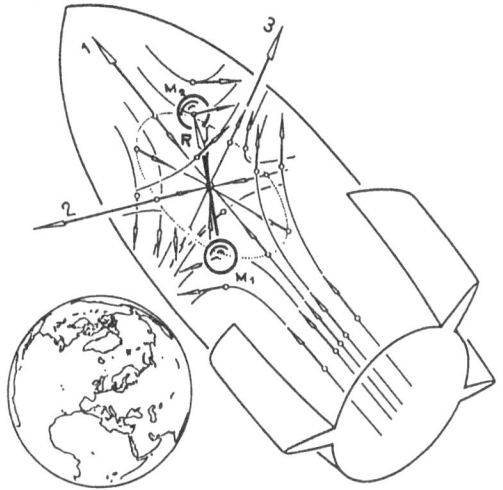

represented by its field lines. The center of mass of the spacecraft is the only acceleration-free point. In all other points accelerometers would measure accelerations different from zero.

The situation can be described mathematically by formulation of the equation of relative motion of the particles M_1 and M_2. Neglecting the gravity interaction between the test masses M_1 and M_2 the equation of motion reads

Fig.1: Field lines of the tidal field

$$\mu \ddot{R} = G \qquad (1)$$

with $\mu = M_1 M_2 / (M_1 + M_2)$ the reduced mass,

$\ddot{R}$ the relative acceleration of the two test masses,

$G = (M_1 K_2 - M_2 K_1)/(M_1 + M_2)$ the tidal force   and

K_1, K_2 the gravity forces acting on the test masses.

The tidal force can be derived from the tidal potential V as following

$$G = \mu \, \nabla_R V \; .$$

In a first linear approximation the tidal potential might be expressed by the Earth's gravity tensor U

$$V = \frac{1}{2} R \cdot U \cdot R$$

so that the tidal force reads

$$G = \mu \, R \cdot U \quad .$$

The gravity tensor U is symmetric and has five independent elements because of the validity of the LAPLACE equation in mass free space

$$\nabla_R \cdot \nabla_R V = \nabla_R \cdot G = \mu \sum_{i=1}^{3} U_{ii} = 0 \, .$$

The local orthogonal triad where the gravity tensor takes diagonal form
is the distinguished triad which can be detected in free fall. It
should be pointed out that the 3-axis of this principal axis triad
(fig.1) deviates from the plumbline direction and both do not coincide
with a line through the geocenter. Gravity gradient stabilization of
satellites can be performed only according to the principal axis triad.
The off-diagonal elements of the gravity tensor decomposed with respect
to a certain, e.g.satellite fixed triad, can be considered as functionals of the angles which describe the misalignment of this triad with
respect to the principle axis triad. One easily sees that the precise
knowledge of the orientation of the local triad is an important pre-
requisite of satellite gravity gradiometry. Indeed, the precise measurement of the orientation of the gradiometer is, besides the high
accelerometer sensitivity, one of the most demanding tasks of such a
future mission. It can be proved easily that misalignment of the local
triad by the small angles, α, β, and γ around the 1,2 and 3-axis produce
of change of gravity tensor components according (Schneider, 1984).

$$
(\bar{U}_{ij}) = \begin{pmatrix} U_{11} & U_{12} + \gamma(U_{22} - U_{11}) & U_{13} + \beta(U_{11} - U_{33}) \\ U_{12} + \gamma(U_{22} - U_{11}) & U_{22} & U_{23} + \alpha(U_{33} - U_{22}) \\ U_{13} + \beta(U_{11} - U_{33}) & U_{23} + \alpha(U_{33} - U_{22}) & U_{33} \end{pmatrix}
$$

With $U_{11} \approx U_{22} \approx 1400$ E.U. and $U_{33} \approx 2800$ E.U. (at altitude 200 km) the
orientation errors in the U_{13} and U_{23} components are magnified by a factor 1400 while the other elements are not affected in a first approximation, in the case that the gradiometer is (approximately) geocenter
oriented.
There are two possibilities to measure the tidal field (Schneider,
1988): In the first indirect possibility the kinematics $R(t)$ of the
relative motion of the two bodies M_1 and M_2 is used to determine the
field parameters modelling the tidal force in equation (1). This idea
is envisaged in the satellite-to-satellite tracking technique. As already mentioned, this technique is applicable in a high-low or a low-
low mode, but it is imaginable also in a micro-mode with test mass separations comparable to those envisaged in gradiometers.

A second possibility is the direct measurement of the tidal field, with
again two alternatives: in the first alternative the relative force K
between two free falling test masses is measured with an in-line oriented dynometer (fig.2). In the second variant the transversal part of
the tidal force, the torque **M** acting on the pair of test particles, is
measured analysing the change of angular momentum (fig.3).

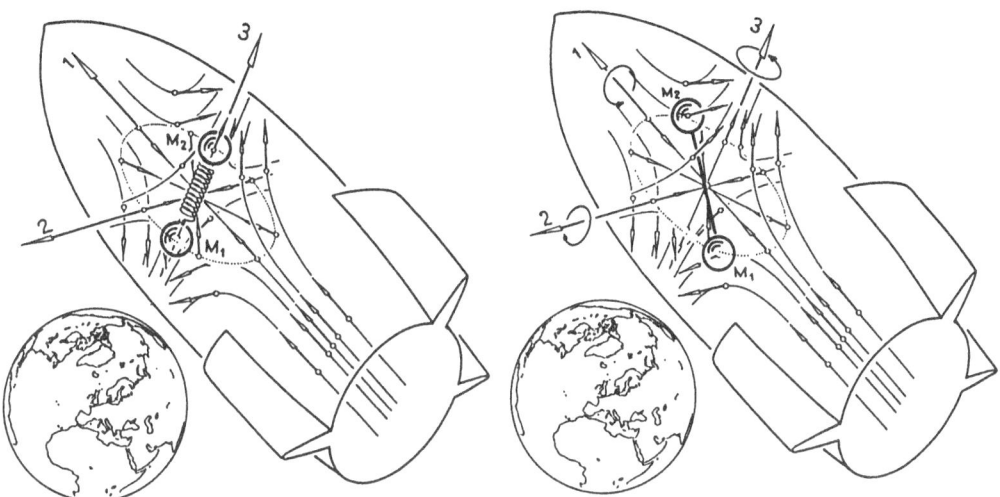

Fig. 2: Principle of the
         in-line gradiometer

Fig.3: Principle of the cross
        component gradiometer

Let us consider, for simplicity, two test particles situated in the
2-3 plane of a local coordinate system, where ϑ describes the angle of
the test mass configuration with respect to the 3-axis. The relative
force in case of an in-line gradiometer reads

$$K = e \cdot G = \mu \frac{R}{2} (U_{22}+U_{33}+ \cos 2\vartheta (U_{33}-U_{22}) + 2 \sin 2\vartheta U_{23})$$

and the torque **M** in case of a cross-component gradiometer

$$M = R \times G = \mu \frac{R^2}{2} \begin{pmatrix} -2 \cos 2\vartheta \ U_{23} + \sin 2\vartheta (U_{33}-U_{22}) \\ \sin 2\vartheta \ U_{12} + (1 + \cos 2\vartheta)U_{13} \\ (\cos 2\vartheta-1)U_{12}- \sin 2\vartheta \ U_{13} \end{pmatrix}$$

The equations of motion read for the relative distance

$$\ddot{R} = \frac{1}{2} (\dot{R}^2 - \dot{R}^2) + \frac{R}{2} (U_{22}+ U_{33} + \cos 2\vartheta(U_{33}-U_{22}) + 2 \sin 2\vartheta U_{23}) \quad (2)$$

and for the rotation (in case of a rigid dumb bell)

$$\dot{d} = \theta^{-1} (M - d \times \theta \cdot d)$$

with the tensor of inertia θ and the rotation vetor **d**. With $\mu = m/2$ and
the moment of inertia in 1-direction, $m R^2/2$, we get for the latter
equation

$$\dot{d}_1 = - \cos 2\vartheta \ U_{23} + \sin 2\vartheta \ \frac{U_{33} - U_{22}}{2} . \quad (3)$$

If the dumb bell is oriented according to the 3-axis and the proof
masses constrained during the measurement procedure the equations
simplify to

$$\ddot{R} = R \, U_{33} \qquad \text{and} \qquad \dot{d}_1 = - U_{23} \; .$$

We see that the measurement of the acceleration $\ddot{R}$ and the angular velocity $\dot{d}_1$ give the tensor components U_{33} and U_{23}.

Based on these equations there are a variety of possibilities to design
a satellite gradiometer. Following Rummel (1989), the design is
determined by three choices

- whether the instrument frame is kept space stable or rotating
  (e.g. earth pointing)
- whether the proof-masses are drifting free or are constrained (and
  in which way) and
- whether the proof-masses are kept in free fall (active drag free
  system) or solely shielded against drag.

Depending on the instrument design a proper modelling of the underlying
physical laws is necessary. The interested reader is refered to e.g.
Rummel (1986,1989).

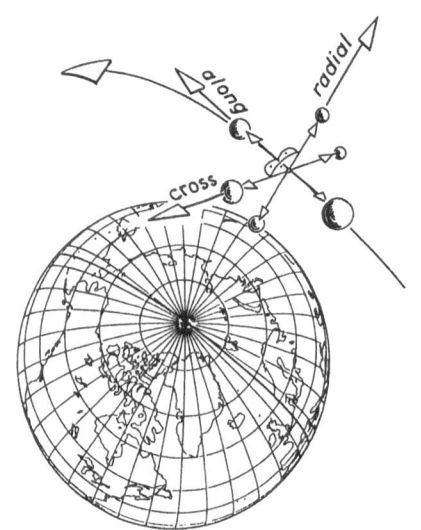

Fig. 4: A simple model of a three
axis "satellite gravity
gradiometer", where the proof-
masses are kept in free fall.

In the following we will consider
a three axis "satellite gravity
gradiometer" where the proof-masses are drifting free and shielded
against non-conservative forces.
The only force ruling the motion
of the test particles shall be the
Earth's gravitation. As measurements, we will only consider in-
line accelerations - so that the
investigations can be based on
equation (2).

3. THE SIGNAL TO BE DETECTED

The purpose of this chapter is to illustrate the size of signals which
gravity gradiometers have to be detected.
A spherical harmonics expansion complete up to degree 180 is considered
to be the pseudo-real gravity field (Rapp, 1981). The reference gravity

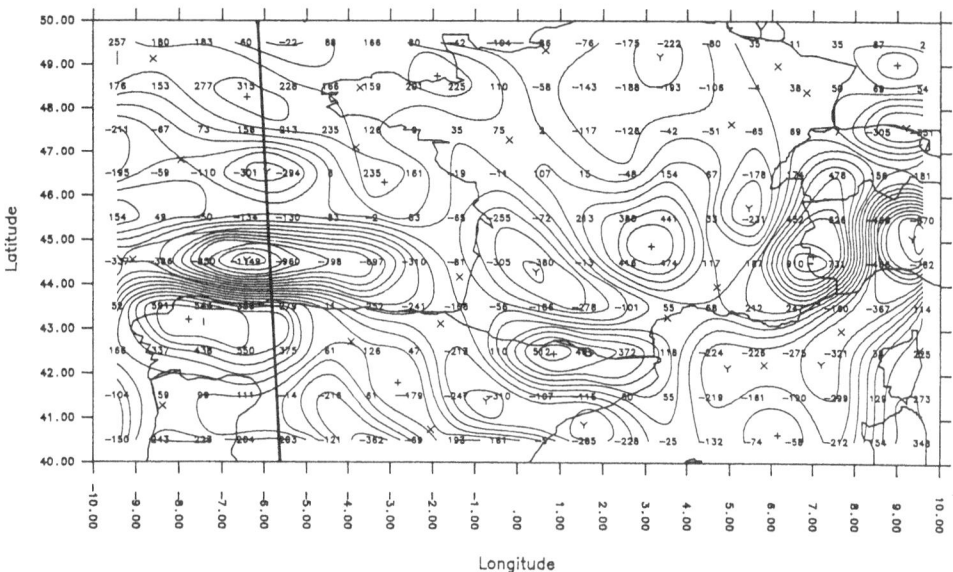

Fig.5: Residual gravity anomalies (in 0.1 mGal) derived from the difference of the pseudo-real gravity field (spherical harmonics expansion complete up to degree 180) and the reference gravity field (spherical harmonics expansion complete up to degree 36); satellite ground track used in table 2.

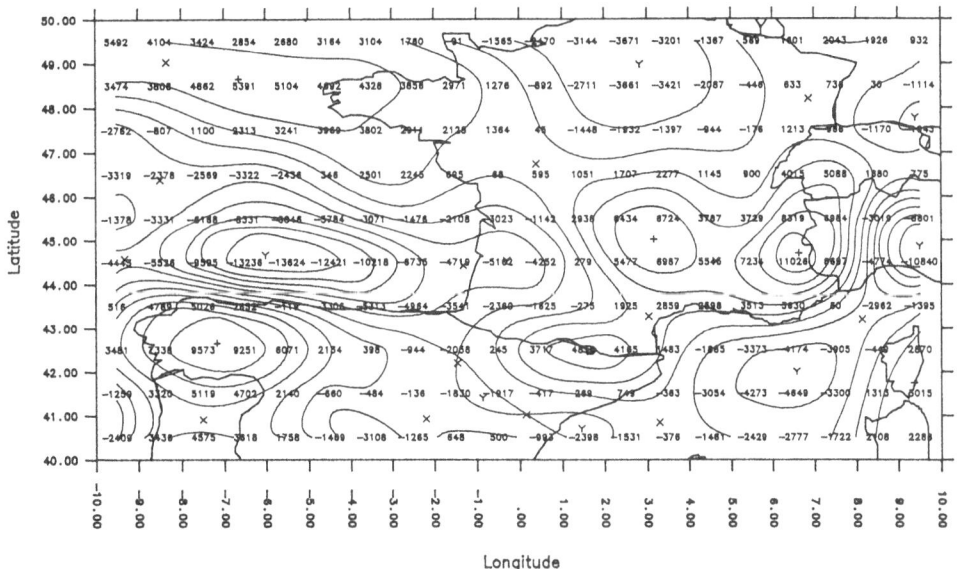

Fig.6: zz-component of the gravity tensor derived from the residual gravity field (fig.5)(in 0.0001 E.U.) in an altitude of 160 km refered to the local astronomical coodinate systems.

field shall be a spherical harmonics expansion complete up to degree 36 of GRIM 3-L1 (Reigber, et al., 1985).

From the residual potential coefficients (pseudo-real minus reference field) gravity anomalies were derived. Fig.5 shows the 1° x 1° residual anomalies for an area, which was adopted for the simulation study in chapter 4, and fig.6 shows the zz-component of the residual gravity tensor in an altitude of 160 km. Table 1 shows the r.m.s.-values for all six tensor components at the three altitudes 160 km, 200 km and 240 km but here refered to the GRS 1980 as reference, see Rummel (1988). This table gives an impression of the relation of the size of

U_{ij} alt	160 km	200 km	240 km
xy	0.12	0.09	0.07
yz	0.25	0.19	0.16
xz	0.26	0.20	0.17
xx	0.21	0.16	0.13
yy	0.22	0.17	0.14
zz	0.37	0.29	0.24

Table 1: r.m.s.-values of the residual gravity tensor (Rummel,1988)

the tensor components and its damping with increasing altitude.

For our simple model of a satellite gravity gradiometer (fig.4) it might be interesting to investigate the relative motion of point masses. Fig.7 shows the relative motion of a configuration of six proof-masses in radial, along and cross direction during one revolution. These motions are nearly exclusively the consequence of the central term of the Earth's gravity field. The effects of the short wavelength parts of the gravity field are depicted in table 2 in terms of relative velocity (groundtrack see fig.5). If an instrumental precision of e.g. 10^{-10} m/sec (equivalent ≈ 0.01 E.U., see table 3) is anticipated one easily recognizes the challenge of such a future mission.

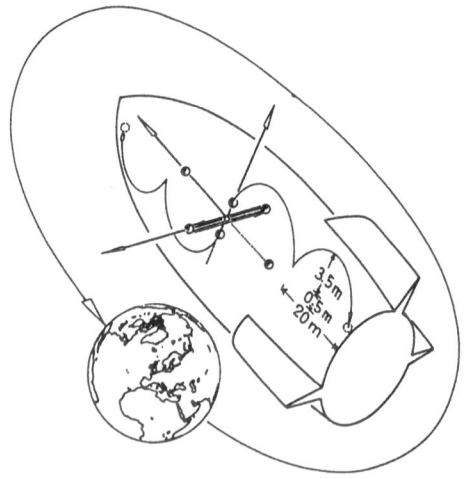

Fig.7: Relative motion of a configuration of six proof-masses during one revolution (Ilk,1989)

spherical harmonics spectral range	r.m.s-relative velocity 10^{-10} m/sec		
	along (1)	cross (2)	radial (3)
60 - 36	28.6	13.7	41.4
90 - 60	17.4	4.2	20.4
120 - 90	5.8	13.2	15.3
150 - 120	2.3	1.9	3.5
180 - 150	1.4	3.9	4.2

Table 2: Gravity field spectral range versus relative velocity

4. SIMULATION STUDY

The size of the signal to be expected in a future gradiometry experiment is very small, especially in the high frequency part of the gravity field spectrum, where it is only a little larger than the observation accuracy itself. Therefore the question arises whether and with what accuracy is it possible to derive gravity field parameters from noisy signal-measurements. The most reliable procedure to get an answer to this question is to perform a simulation study where the aspects to be investigated are modelled realistically.

A global gravity parameter recovery study is a very demanding task both with respect to computation time and memory capacity. Therefore, the simulation study is restricted to a regional gravity field recovery experiment related to the test area depicted in figures 5 and 6 and to the gravity fields described in chapter 3. As spacecraft carrying the gradiometer a low altitude satellite in near polar orbit (i $\approx$ 96°) was selected with altitudes discussed in connection with the Aristoteles mission. Four different altitudes are envisaged: 160 km, 200 km, 210 km and 240 km. For these cases fig.8 shows the groundtracks for a mission period of 100 days. A simple model of a three axis satellite gravity gradiometer, where the proof-masses are kept in free fall, was considered (fig.4). As proof mass separations three cases were simulated: 60 cm, 90 cm and 120 cm.

Most of the gravity gradiometers proposed are based on the "direct" measurement principle, e.g. relative accelerations are measured. The present simulation study is based on the "indirect" measurement method. As observation type range-rates are used-which show a slightly different spectral

$\begin{array}{c}m_{\dot{R}}\\R\end{array}$ cm	10^{-10} m/sec		
	1	2	3
60	1.6	3.1	4.7
90	1.0	2.1	3.1
120	0.8	1.6	2.4

Table 3: Accuracy of the "free fall gradiometer" in terms of 0.01 E.U. in dependence of test mass separation R and range-rate measurement accuracy $m_{\dot{R}}$

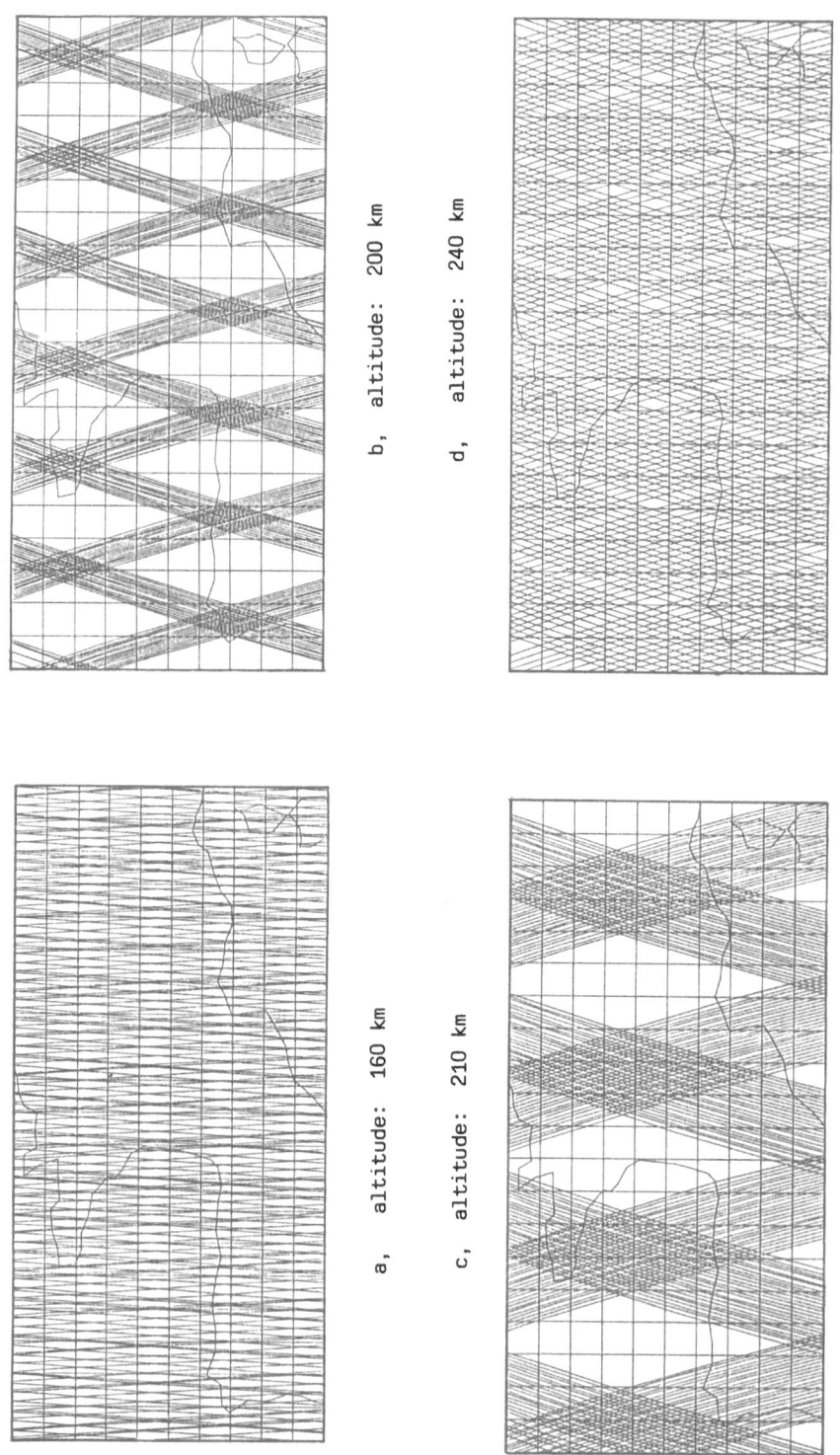

b, altitude: 200 km

d, altitude: 240 km

a, altitude: 160 km

c, altitude: 210 km

Fig.8: Ground tracks of low altitude satellites for a mission period of 90 days over the region $\varphi = [40^o, 50^o]$, $\lambda = [-10^o, 10^o]$

sensitivity. In case of short arcs this difference is of marginal importance - so the results can be considered to be representative for any kind of gradiometer (see the conversion table 3).

The technique applied for the regional mapping of the gravity field is based on the solution of a boundary value problem associated with a system of ordinary differential equations describing the relative motion of two satellites (Ilk, 1986)

The method can be characterised by the following steps:

- derivation of the range-rates as function of the unknown field parameters by measurement of range-rates along the short arcs over the test area and reduction by the reference part
- determination of the reduced observables by FOURIER expansion of the range-rate function applying a numerical quadrature technique
- regularised solution of an overdetermined system of linear equations describing the downward continuation process of gravity field functionals from satellite altitude to the Earth's surface where the local gravity field parameters are defined.

The main purpose of this study is to investigate the effect of different observation error levels on the recovery results. For this task the simulated range-rates are contaminated with normally distributed random errors at the observation points. It is anticipated that the observations are taken at the nodes of the quadrature formula, which are not equidistant. The mean data density is approximately one observation per 3 seconds.

The 200 recovered residual gravity anomalies $\hat{g}$ were compared with the simulated (error-free) values g and root mean square (rms) values derived

$$\text{rms:} = \sqrt{\|g-\hat{g}\|_2^2/200} = \sqrt{\sum_{i=1}^{200}(g_i-\hat{g}_i)^2/200} \quad .$$

The figures 9a to 9f show the recovery results in terms of rms values (mGal) as a function of different range-rate error levels and for all combinations (one-, two-, three axis gradiometer). Table 4 shows the recovery accuracy as a function of the mission duration for selected test examples.

5. DISCUSSION OF THE RESULTS AND CONCLUSIONS

Figures 9a to 9f clearly show the benefit of a three-component gradiometer over two-component or, even more pronounced, over one component gradiometers. The poor results of the cross component test mass

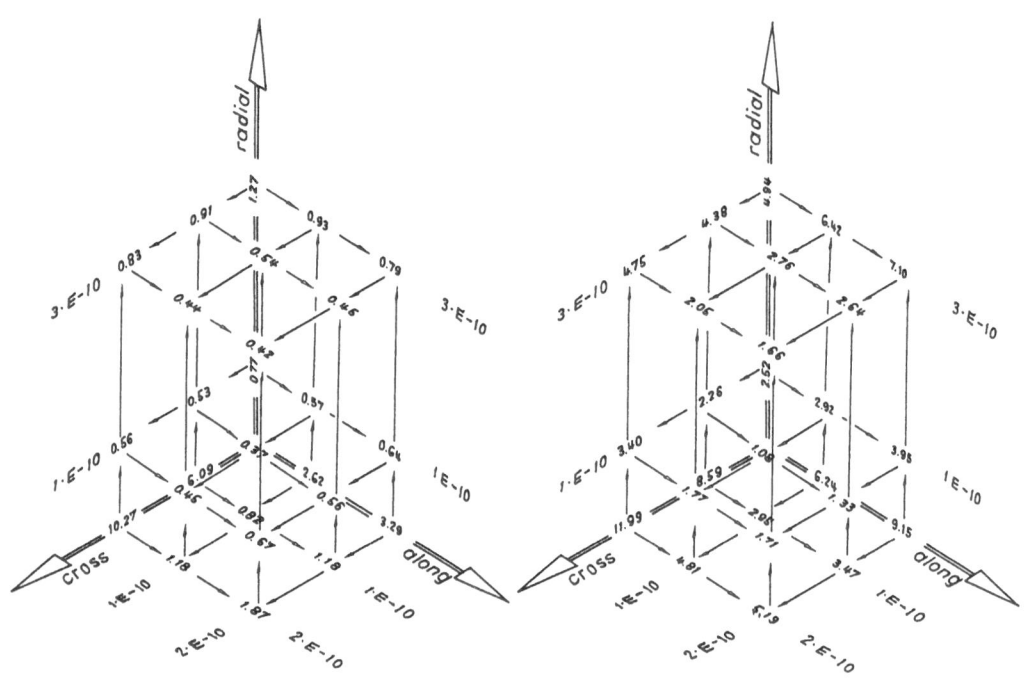

a, altitude 160 km, R = 60 cm

b, altitude 200 km, R = 60 cm

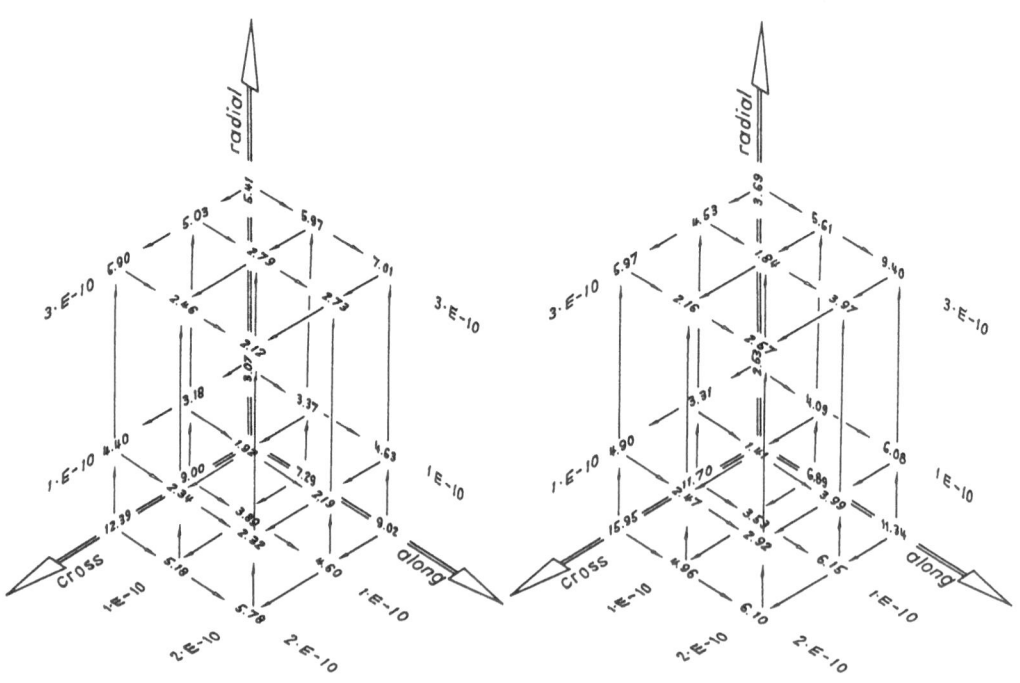

c, altitude 210 km, R = 60 cm

d, altitude 240 km, R = 60 cm

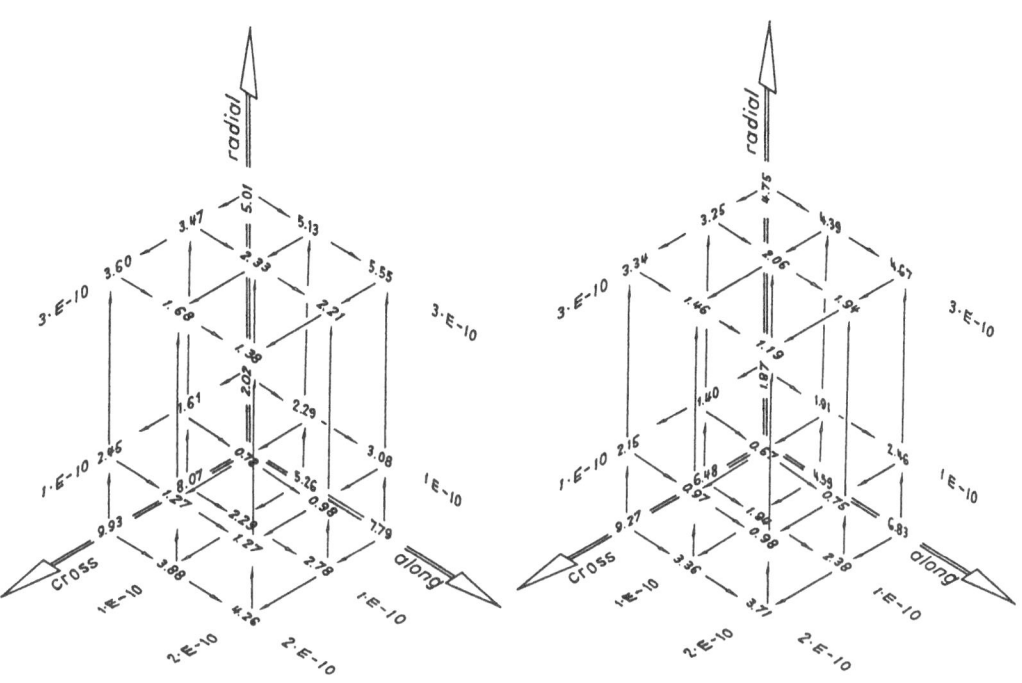

e, altitude 200 km, R = 90 cm   f, altitude 200 km, R = 120 cm

Fig.9 Recovery results in terms of rms-values (mGal) depending on
       different range-rate error levels and for all combinations (one-,
       two-, three axis gradiometer); mission period 90 days; 182 short
       arcs used; approximate number of observations used: one component
       gradiometer: 14 500, two-component gradiometer: 29 000,
       three component gradiometer: 43 500.

mission duration			rms-values (in mGal)			
days	arcs	observ.	160 km	200 km	210 km	240 km
90	182	43 680	0.35	1.08	1.92	1.41
60	121	29 040	0.57	1.50	2.83	1.56
45	91	21 840	0.63	1.77	3.28	1.81
30	61	14 640	0.76	2.23	4.10	2.23
15	31	7 440	0.95	3.35	5.74	3.29
8	16	3 840	1.34	4.76	7.30	5.38
4	8	1 920	9.28	9.55	9.11	11.40

Table 4: Gravity field recovery results in dependence of the mission
          duration (three-axis gradiometer, test mass separation
          60 cm, observation accuracy 0.016 E.U.)

configuration are in coincidence with the signals given in table 2.
It is very unlikely that this phenomenon is valid for all regions all
over the Earth because generally the signals are similar in local horizontal directions. On the other hand, the fact that a radial test mass
configuration gives the best results seems to be a general valid phenomenon because of the largest signal in this direction. The recovery
quality in case of two-component gradiometers is approximately identical for all combinations, though it is somewhat surprising that the
combination of the radial component with the cross component gives
slightly better results than the radial-along combination. It is also
interesting to note that the along-cross combination gives only worse
recovering results in case of low satellite altitudes (160 km) but not
in higher ones (see e.g. 240 km).

The comparison of figures 9b,c with 9a,d show that a good ground track
coverage is an important aspect of the design of a gradiometry mission,
which is already pointed out in Reigber et al.,(1989). In this study an
optimal altitude of 198.91 km is proposed. Obviously a satellite with
an altitude of 240 km and a good ground track coverage (fig.9d) gives
comparable results as in the case of a lower altitude satellite
(200 km) with poor coverage (fig.9b) except those combinations with
high observation accuracy.

The comparison of figures 9b,e,and f show that it is advantageous to
choose a test mass separation as large as possible. There is no linear
improvement; the benefit seems to be larger in case of higher observation accuracies.

As expected, the recovery accuracies are strongly dependent on the mission duration (table 4). Remarkable is the fact that after already two
weeks a reasonable result can be obtained and after 90 days observation
acquisition the results can be improved only marginally.

At last we would like to remark that the numbers depicted in figures
9a to 9f, are derived by differencing true values and gravity anomalies recovered from noisy observations. As in any real observation
campaign the derived quantities may better coincide with the true
values in case of worse instrumentation accuracy than in the case of
better ones. Additionally, we have to keep in mind that downward
continuation is an unstable process and that regularisation can remove
the unstabilities only partially. This is the reason that some of the
recovery results are slightly different from what we would expect when
a higher instrumentation accuracy is applied.

In this study only the effect of observation inaccuracies on the recovered anomalies was investigated. We did not investigate the effect of orbit and attitude errors. In the case of in-line measurements attitude errors may affect the recovery results at marginal amounts as already mentioned in chapter 2 provided that the instrument is oriented according to the local astronomical system. The orbit determination errors have a minor influence on the recovery results, at least when the absolute position error is considered (for further details see e.g. (Ilk, 1988).

In conclusion one can state that satellite gravity gradiometry gives satisfactory results if the instrument is gravity stabilised and in-line components are used. Important aspects of a mission design are orbit tuning to reach a regular ground track coverage and to apply a three component gradiometer with test mass separations as large as possible. Orbit and attitude errors may have minor influence an the recovery results - but this point has to be investigated in more detail.

REFERENCES

BALMINO, G., BARLIER, F., BERNARD, A., BONZAT, C. RUMMEL, R., TOUBOUL, P. (1985) Proposal for a sat. gravity gradiometer experiment for the geosciences

ILK, K.H. (1986) On the regional mapping of gravitation with two satellites, proceedings of the I. Hotine-Marussi-Symposium on Mathematical Geodesy, Roma, 3-6 June 1985, Vol. 2, pp 807-832, Milano

ILK, K.H. (1988) Regional Gravity Field Mapping: Satellite Gravity Gradiometry versus Satellite-to-Satellite Tracking Techniques, Proc.of the IUGG General Assembly in Vancouver, IAG, Vol.II, pp. 284, Paris

ILK, K.H. (1989) Zukünftige Möglichkeiten der globalen hochauflösenden Schwerefeldbestimmung, in "Satellitengeodäsie: Abschlußbericht des Sonderforschungsbereiches 78 an der Techn.Univ. München 1970-1986/DGG", Dt. Forschungsgemeinschaft, Hrsg.v.M. Schneider, VCH Weinheim

KEATING, T., TAYLER, P., KAHN, W., LERCH, F. (1986) Geopotential Research Mission science, engineering and program summary, NASA Tech. Mem. 86240

LERCH, F.J., PUTNEY, B.H., WAGNER, C.A., KLOSKO, S.M. (1981) Goddard Earth Models for Oceanographic Applications (GEM 10B and 10C), Marine Geodesy, 5,2, 145-187

MARSH, J.G., LERCH, F.J., PUTNEY, B.H., CHRISTODOULIDIS, D.C., FELSEN-TREGER, T.L., SANCHEZ, B.V., SMITH, D.E., KLOSKO, S.M., MARTIN, T.V., PAVLIS, E.C., ROBBINS, J.M., WILLIAMSON, R.G., COLOMBO, O.L., CHANDLER, N.L., RACHLIN, K.E., PATEL, G.B., BHATI, S. and CHINN, D.S. (1987) An Improved Model of the Earth's Gravitational Field: GEM-T1, NASA Technical Memorandum 4019

PAIK, H.J., LEUNG, J.S., MORGAN, S.H., PARKER, J. (1988) Global Gravity Survey by an Orbiting Gravity Gradiometer, EOS, Nov. 29

RAPP, R.H. (1986) Global Geopotential Solutions, Lecture Notes in Earth Sciences, Vol. 7, Mathematical and Numerical Techniques in Physical Geodesy, Edited by H. Sünkel, Springer-Verlag Berlin-Heidelberg

REIGBER, Ch., BALMINO, G., MÜLLER, H., BOSCH, W. and MOYNOT, B. (1985) GRIM gravity model improvement using LAGEOS (GRIM3-L1), 7. Geophys. Res., Vol. 90, No. B11, pp. 9285-9299, Sept.30

REIGBER; Ch., DROZYNER, A., BODE, A. (1989) Orbit and grad^2T generation, in "Study on precise gravity field determination methods and mission requirements", ESA Contract No. 7521/87/F/FL, Final Report, Paris

RUMMEL, R.: Satellite Gradiometry (1986) in: Lecture Notes in Earth Sciences, Vol. 7, 317-363, Springer, Berlin

RUMMEL, R. (1989) SGG principles and state of the art: instruments proposed, simulation performed, emphasis on differential accelerometry, in "Study on precise gravity field determination methods and mission requirements", ESA Contract No. 7521/87/F/FL, Final Report, Paris

SCHNEIDER, M. (1984) Himmelsmechanik, BI-Wissenschaftsverlag Mannheim, Nachdr.d.2.Aufl.

SCHNEIDER, M. (1988) Satellitengeodäsie, BI-Wissenschaftsverlag, Mannheim

SPACEBORNE GRAVITY GRADIOMETERS (1983) Proc. workshop held at NASA Goddard Space Flight Center, Greenbelt Md.

Separation of Inertia and Gravitation in Airborne Gravimetry with GPS

A. Kleusberg
Geodetic Research Laboratory
University of New Brunswick
CANADA

ABSTRACT:

Gravimeters on moving platforms measure the sum of gravitational and inertial acceleration. To relate these measurements to the gravity field of the earth, the accelerations caused by platform motion must be separated from the gravity signal. This separation constitutes one of the major problems in the proper analysis of airborne gravity measurements. In principle, the separation is possible by observing or modelling the inertial accelerations of the platform simultaneously and independently. Observing precise inertial accelerations directly has become possible by analysing the second time derivative of range measurements between the platform and the satellites of the Global Positioning System (GPS). The paper reviews the accuracy requirements for inertial accelerations in typical airborne gravimetry applications, and describes the principle of deriving accelerations from carrier phase observations. Particular attention is given to the spectral distribution of the platform acceleration errors, and their relation to the gravity signal spectral pass band.

1. INTRODUCTION

Geodesists measure gravity for the purpose of determining the geoid and the external gravity field of the earth. Exploration geophysicists measure gravity for the purpose of interpreting gravity anomalies in terms of mass anomalies. In either case, gravity data acquisition is a laborious and time consuming process. Over land areas, gravity data is usually collected statically at discrete points. Marine gravity is collected onboard survey vessels by integrating measured vertical acceleration over sufficiently long periods to ensure the elimination of vertical platform accelerations caused by heave motion of the vessel.

Airborne gravimetry promises to become a considerably faster and more economical tool of gravity determination. However, an efficient way is required of extracting the gravity signal from the accelerations measured on a moving platform. Obviously, for an airborne platform, this separation cannot be accomplished just by integrating and averaging as in marine gravimetry. Vertical platform accelerations have to be provided from external sources. The present paper explores the possibility of determining these platform accelerations using Global Positioning System (GPS) measurements.

2. AIRBORNE GRAVIMETRY

We restrict the following discussion to reconnaissance airborne gravity surveys as specified in Bower and Halpenny (1986) for fixed-wing aircraft. These specifications call for an accuracy of mean gravity values of about one mGal ($1 \cdot 10^{-5}$ m/s^2) with a spatial resolution of 5 km. For a fixed-wing aircraft with a nominal cruising speed of 300 km/h, the resulting gravimeter integration interval is 120 seconds. Therefore, low pass filtering with a cut-off frequency of 0.0083 Hz can be applied to raw measurements.

2.1 Accelerations In A Moving Reference Frame

Any gravimeter (accelerometer) on a moving platform senses the sum of gravitational and inertial acceleration. Any distortions resulting from other forces will not be considered here. Thus, in inertial space the total vectorial acceleration $\mathbf{a}$ can be represented by

$$\mathbf{a} \;=\; \tilde{\mathbf{g}} \,-\, d_{tt}\,\mathbf{x} \tag{1}$$

where $\tilde{\mathbf{g}}$ is the vector of gravitational acceleration, $\mathbf{x}$ is the platform position vector, and d_{tt} is the second time derivative in inertial space. Transforming equation (1) to an earth-fixed coordinate system rotating with rotation velocity vector $\mathbf{w}$ with respect to inertial space yields

$$\mathbf{a} = \tilde{\mathbf{g}} - \ddot{\mathbf{x}} - \mathbf{w} \wedge \mathbf{w} \wedge \mathbf{x} - 2\,\mathbf{w} \wedge \dot{\mathbf{x}} - \dot{\mathbf{w}} \wedge \mathbf{x} \tag{2}$$

where $\mathbf{a}$, $\tilde{\mathbf{g}}$, and $\mathbf{x}$ are as defined before, and a dot above a vector denotes the time derivative in the rotating earth fixed coordinate system. The symbol '$\wedge$' denotes the vector product. The third, fourth, and fifth terms on the right side of eqn.(2) are known as centrifugal, Coriolis, and Euler acceleration respectively. The centrifugal acceleration sums up with the gravitational acceleration $\tilde{\mathbf{g}}$ to yield gravity $\mathbf{g}$ according to

$$\mathbf{g} = \tilde{\mathbf{g}} - \mathbf{w} \wedge \mathbf{w} \wedge \mathbf{x}. \tag{3}$$

Since variations in the rotation vector are negligible in the present context, $\mathbf{w}$ can be assumed to be constant and parallel to the z-axis of the earth-fixed coordinate system. Therefore, the Euler acceleration will be omitted in the following derivations, and denoting the Coriolis acceleration by $\mathbf{C}$, we obtain from eqn.(2) the simplified form

$$\mathbf{a} = \mathbf{g} - \ddot{\mathbf{x}} - \mathbf{C}. \tag{4}$$

Gravimeters mounted on moving platforms are primarily sensitive along a single axis, which is kept aligned to the local vertical by appropriate mechanisms (e.g. Brozena et al., 1986). For a more adequate modelling of the resulting gravimeter observations, we transform eqn. (4) into the local geodetic (ellipsoidal) coordinate system with orthogonal base vectors e_ϕ, e_λ, e_h pointing towards north, east, and ellipsoidal normal respectively. Denoting the platform velocity components along these axes by

$$\begin{aligned} v_\phi &= (M+h)\,\dot{\phi} \\ v_\lambda &= (N+h)\,\cos\phi\,\dot{\lambda} \\ v_h &= \dot{h} \end{aligned} \tag{5}$$

where ϕ, λ, h are geodetic coordinates, and M and N are the principle radii of curvature in meridian and prime vertical direction respectively (e.g. Bomford, 1971), we obtain

$$\ddot{x} = \left\{ \frac{\tan\phi}{N+h} v_\lambda v_\lambda + \frac{1}{M+h} v_h v_\phi + \dot{v}_\phi \right\} e_\phi$$

$$+ \left\{ -\frac{\tan\phi}{N+h} v_\lambda v_\phi + \frac{1}{N+h} v_h v_\lambda + \dot{v}_\lambda \right\} e_\lambda$$

$$+ \left\{ -\frac{1}{M+h} v_\phi v_\phi - \frac{1}{N+h} v_\lambda v_\lambda + \dot{v}_h \right\} e_h \quad (6)$$

and

$$C = 2 w \left[\{v_\lambda \sin\phi\} e_\phi + \{-v_\phi \sin\phi + v_h \cos\phi\} e_\lambda + \{-v_\lambda \cos\phi\} e_h \right] \quad (7)$$

where w is the length of the rotation vector **w**. From eqns. (4), (6), and (7) we obtain for the total acceleration components along the local geodetic system

$$a_\phi = g_\phi - \frac{\tan\phi}{N+h} v_\lambda v_\lambda - \frac{1}{M+h} v_h v_\phi - 2 w v_\lambda \sin\phi - \dot{v}_\phi \quad (8)$$

$$a_\lambda = g_\lambda + \frac{\tan\phi}{N+h} v_\lambda v_\phi - \frac{1}{N+h} v_h v_\lambda + 2 w (v_\phi \sin\phi - v_h \cos\phi) - \dot{v}_\lambda \quad (9)$$

$$a_h = g_h + \frac{1}{M+h} v_\phi v_\phi + \frac{1}{N+h} v_\lambda v_\lambda + 2 w v_\lambda \cos\phi - \dot{v}_h . \quad (10)$$

2.2 Approximations

As mentioned above, we consider in this study the potential of airborne gravimetry for gravity surveys with an accuracy of one mGal. In order to simplify equations (8) through (10) accordingly, we assess the significance of the terms on the right hand sides. We base this assessment on a nominal horizontal aircraft speed of 300 km/h, zero nominal vertical velocity, and 45 degrees of latitude. Neglecting errors less than 0.5 mGal, we can replace the principle radii of curvature, N and M, by a mean curvature, $R = \sqrt{NM}$, and substitute the scalar gravity value

$$g = \sqrt{g_\phi \cdot g_\phi + g_\lambda \cdot g_\lambda + g_h \cdot g_h} \quad (11)$$

for the negative of the outward vertical gravity component g_h. Thus we obtain from eqn.(10)

$$a_h = -g + E_h - \dot{v}_h \quad (12)$$

with

$$E_h = \frac{1}{R+h} (v_\phi v_\phi + v_\lambda v_\lambda) + 2 w v_\lambda \cos\phi . \quad (13)$$

Similarly we obtain for the horizontal accelerations

$$a_\phi = g_\phi + E_\phi - \dot{v}_\phi \tag{14}$$

$$a_\lambda = g_\lambda + E_\lambda - \dot{v}_\lambda \tag{15}$$

with

$$E_\phi = -\frac{\tan\phi}{R+h} v_\lambda v_\lambda - 2 w v_\lambda \sin\phi \tag{16}$$

$$E_\lambda = \frac{\tan\phi}{R+h} v_\lambda v_\phi + 2 w v_\phi \sin\phi. \tag{17}$$

The Eötvös correction, E_h, is of the order of 1000 mGal for east-west flights and of the order of 100 mGal for north-south flights. E_ϕ is of the order of 1000 mGal for east-west flights and vanishes for north-south flights whereas E_λ is zero for east-west flights and of the order of 1000 mGal for north-south flights.

2.3 Airborne Gravimeter Platform Measurements

To utilise the acceleration measurements along the sensitive axis of the gravimeter for gravity field information, this axis has to be kept aligned to a well defined direction and the non-gravitational accelerations along the sensitive axis have to be provided by external measurements. We assume in the following that nominally the preferred direction for the gravimeter axis is the ellipsoidal vertical, e_h. The platform stabilisation is assumed to be achieved through gyroscopes controlled by auxiliary accelerometers mounted orthogonally to the gravimeter axis (cf. LaCoste, 1967). The sensitive axes of the gravimeter and the two auxiliary accelerometers constitute an orthogonal measurement frame e_1, e_2, e_3 which is rotated by angles α, β about the e_ϕ, e_λ axes respectively. Because of the aforementioned platform control mechanism, the angles α and β remain small and the approximations

$$\sin \alpha = \alpha, \ \cos \alpha = 1 - \alpha^2/2, \ \sin \beta = \beta, \ \cos \beta = 1 - \beta^2/2 \tag{18}$$

can be used. Denoting the accelerometer measurements in the e_1, e_2, e_3 frame by b_1, b_2, b_3, we obtain

$$b_1 = a_\phi (1-\beta^2/2) - \beta a_h \tag{19}$$

$$b_2 = a_\lambda (1-\alpha^2/2) + \alpha a_h \tag{20}$$

$$b_3 = a_\phi \beta - a_\lambda \alpha + a_h (1-\alpha^2/2-\beta^2/2). \tag{21}$$

To keep errors due to non-verticality in the last term of eqn. (21) below 0.5 mGal, the allowable total tilt of the gravimeter axis has to remain within ± 1 mrad (3.5 minutes of arc).

If the platform verticality is controlled by measuring accelerations perpendicular to the (nominally vertical) platform axis, the allowable tilt of ± 1 mrad requires a horizontal acceleration measurement accuracy of about 1 Gal. At this accuracy level, all but the last terms on the right hand side of eqns. (8) and (9) can be neglected to obtain

$$a_\phi = - \dot{v}_\phi \tag{22}$$
$$a_\lambda = - \dot{v}_\lambda . \tag{23}$$

Replacing the horizontal acceleration components in eqns. (19) and (20) by eqns. (22) and (23), and the vertical acceleration by gravity, we obtain the following approximations

$$\beta = - (b_1 + \dot{v}_\phi) / g \tag{24}$$
$$\alpha = (b_2 + \dot{v}_\lambda) / g. \tag{25}$$

These equations describe the platform tilt angles α and β in terms of accelerometer measurements, horizontal platform velocity rates, and gravity. To satisfy the aforementioned 1 mrad accuracy requirement for the tilt angles, b_1, b_2, $\dot{v}_\phi$ and $\dot{v}_\lambda$ have to be measured with 1 Gal accuracy, and a nominal gravity value can substituted for g.

Accelerometer measurements b_3 along the nominally vertical platform axis are the primary source for gravity information. Assuming constant horizontal velocity while measuring b_3 we can omit the first two terms on the right side of eqn. (21), and using eqns. (12) and (13) we obtain

$$g = - b_3 + \frac{1}{R+h} (v_\phi v_\phi + v_\lambda v_\lambda) + 2 w v_\lambda \cos\phi - \dot{v}_h. \tag{26}$$

As mentioned at the beginning of this section, we are interested in the determination of mean gravity values for integration intervals of 120 s. This integration yields

$$\left[\, g \, \right] = - \left[\, b_3 \, \right] + \left[\frac{1}{R+h} (v_\phi v_\phi + v_\lambda v_\lambda) + 2 w v_\lambda \cos\phi \right] - \left[\, \dot{v}_h \, \right] \tag{27}$$

where we have used the square brackets to indicate averaging over the integration interval. To obtain mean gravity with 1 mGal accuracy as specified above, obviously all right hand terms of eqn. (21) have to be determined with comparable accuracy. This requires 0.1 m/sec accuracy for the horizontal platform velocity components in the computation of the Eötvös correction, and 1 mGal accuracy for mean vertical platform accelerations over 120 s. To relate the mean gravity values determined at aircraft altitude to the gravity field of the earth, the position of the gravimeter platform must be known at the several metre accuracy level (assuming a vertical gravity gradient of about -0.3 mGal/m).

2.4 Conventional Positioning And Data Reduction

Conventionally, horizontal positions of the gravimeter platform are determined from terrestrial radio positioning systems. Accurate horizontal platform accelerations to be used for platform verticality control (cf. eqns. (24) and (25)) cannot be obtained from these systems. Therefore, the horizontal accelerometer output is not utilised for platform control in periods of large horizontal accelerations (cf. Brozena et al., 1986). Vertical positions are obtained from altitude measurements with radar altimeters (over open water surfaces) or from atmospheric pressure transducers. Vertical accelerations to be used in eqn. (27) are determined from the second derivative of the altitude measurements. This type of instrumentation suffers from problems regarding accuracy, reliability, and calibration as outlined by Bell et al. (1986). In addition, the use of a multitude of different types of instrumentation for different parts of the data acquisition and reduction task creates a highly complex overall system (cf. Brozena et al., 1986). Replacing all positioning systems with a single, simpler and more accurate sensor seems to be highly desirable.

3. THE RÔLE OF GPS

The Global Positioning System (GPS) is a satellite based radio positioning system designed for high precision real time navigation. Although the system is not yet completed, GPS measurements have been shown to yield superb results in many positioning related applications (cf. Wells et al. (1986)).

The GPS satellite orbits represent the mechanisation of a coordinate system. We shall assume, that the orbit description in terms of ephemerides is given in an earth-fixed reference system. A GPS receiver provides measurements between the receiver and the GPS satellites, which can be used to relate the receiver position to the satellite orbits, and thus determine the receiver position in the earth-fixed reference system.

3.1 GPS Observables

The two basic GPS observation types are pseudoranges P and carrier phases Φ, which can be modelled according to

$$P = \rho + c\,dt + \varepsilon_P \tag{28}$$

and

$$\Phi = \rho + c\,dt + \lambda N + \varepsilon_\Phi \tag{29}$$

where ρ is the geometric range between satellite and receiver, c is the speed of light, dt is the unknown receiver clock offset, λ is the GPS carrier signal wavelength, N is an unknown integer number called carrier phase ambiguity, and ε_P and ε_Φ are pseudorange and carrier phase measurement errors respectively. For the sake of clarity, any terms pertaining to systematic effects and biases have been omitted in the observation equations. For a more complete discussion of the observation equations we refer to Wells et al. (1986). The geometric range ρ depends implicitly on the receiver position and the satellite position. Assuming known satellite orbits, four or more simultaneously observed pseudoranges can be utilised to determine instantaneous receiver position and clock offset. Carrier phases observed simultaneously on several satellite signals over extended periods of time are used in static positioning to determine simultaneously receiver position and carrier phase ambiguity.

3.2 GPS Relative Positioning Accuracy

Because of the aforementioned systematic effects and biases in GPS observations, absolute positions determined with GPS measurements are limited in accuracy to several tens of metres. Since most of the accuracy limiting systematic effects have high spatial correlation, more accurate results are obtained for relative positions between simultaneously observing receivers. In this case of GPS relative positioning, relative coordinate accuracies at or below one part per million of baseline length are obtained for static surveying with carrier phase measurements.

In kinematic relative positioning with GPS, position accuracies depend to a large extend on the type of measurement utilised. Typical accuracies for relative kinematic positions obtained from pseudoranges alone are of the order of ten metres (e.g. Lachapelle et al., 1984), and are limited by the pseudorange measurement noise. Combining pseudoranges and carrier phases in the positioning algorithm reduces the impact of pseudorange noise tremendously. This technique has been shown to produce kinematic relative position accuracies below the one metre level (eg. Cannon, 1987; Kleusberg, 1986), and using a simple model for the platform dynamics yields platform velocity estimates better than 0.1 metre per second (e.g. Schwarz et al., 1987). If carrier phase ambiguities (cf. eqn. (29)) can be determined and fixed to their integer values, and if all carrier phase discontinuities can be resolved, kinematic relative position accuracies at the several centimetre level can be obtained (e.g. Mader, 1986; Landau, 1989). However, this accuracy level has been reported so far only for carefully designed experiments and is not available on an operational basis.

4. PLATFORM ACCELERATIONS FROM GPS

As shown in eqns. (28) and (29), GPS is basically a ranging system, and as such relates primarily receiver positions to satellite positions. For a moving receiver, the sequence of range measurements or the sequence of positions represent the platform trajectory as a function of time, and therefore, contain information about the platform kinematics. This section explores different ways to extract platform acceleration from GPS measurements and assesses the related accuracy potential of GPS.

4.1 Obtaining Platform Accelerations From Carrier Phases

The description of airborne gravimeter measurements in section 2.3 and the data reduction process in section 2.4 explained the need for external measurements of the gravimeter platform acceleration. For direct use in eqns. (24) through (27) these accelerations should be provided in terms of geodetic coordinates ϕ, λ, h. The different ways of deriving accelerations from observed carrier phases are depicted schematically in Fig. 1. This figure shows on the left the observed ranges (carrier phases) and their time derivatives, and on the right the platform position, velocity, and acceleration. The differentiation operator is denoted by d_t, and the connection between observation and position related parameters is given by least squares adjustments abbreviated LSQ#1, LSQ#2, LSQ#3. Obviously, there exist several ways to proceed from observed carrier phases to platform accelerations.

The most direct approach consists of computing ϕ, λ, h position time series from carrier phase time series (LSQ#1) and taking the second time derivative of the positions. Obviously, this procedure requires very accurate position determination. To obtain positions at the required accuracy level, two presently non-trivial problems have to be overcome. First, the carrier phase ambiguities N (cf. eqn. (29)) have to be explicitly determined and fixed to their integer value. Secondly, any carrier phase discontinuities (cycle slips) have to be detected and corrected before the position determination can proceed. Despite these problems, results have been obtained with this procedure through careful error correction and elaborate data processing (Brozena et al., 1989).

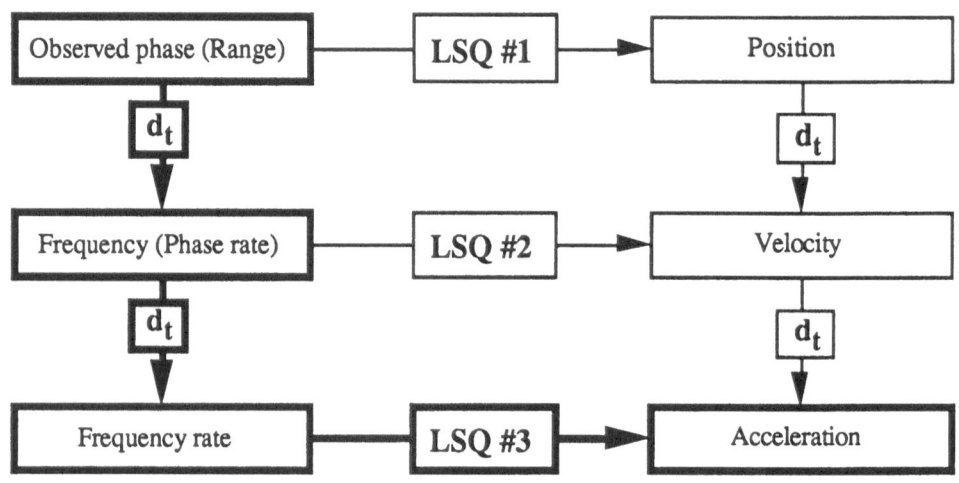

Figure 1: From ranges to platform accelerations

Another way of deriving platform accelerations from GPS carrier phase observations is indicated by bold lines in Fig. 1. It consists of first taking the second time derivative in the observation domain, and then converting the resulting frequency rates to platform accelerations $\dot{v}_\phi$, $\dot{v}_\lambda$, and $\dot{v}_h$ in LSQ#3. There are two main advantages in this procedure. First, the carrier phase ambiguities do not have to be determined explicitly. Being constants (in time), they drop out in the differentiation process. Secondly, cycle slips do not have to be corrected in the carrier phase observations. Detecting cycle slips and discarding the contaminated time derivative is sufficient to eliminate the effect of phase discontinuities. Accuracy requirements for platform positions and velocities in LSQ#3 are at the several metre and several decimetre per second level respectively, and can be easily satisfied by conventional kinematic GPS data processing. It appears that this process is simpler than using LSQ#1 as described above.

4.2 Error Propagation In Numerical Differentiation

Going either of the two ways for the computation of platform accelerations as described in the previous section involves two numerical differentiations. The discrete time series to be differentiated with respect to time are either positions or carrier phase measurements. In general, both time series will be contaminated by errors resulting from measurement noise and unmodelled effects. To understand the impact of these errors on estimated accelerations, we find it useful to recall some aspects of error propagation in numerical differentiation. The important aspects in the present context can be most conveniently shown in the frequency domain. Therefore, in this section and the following one, time series of measurement errors will be described in terms of their power spectral density (PSD) distribution (e.g. Otnes and Enochson, 1978).

If $PSD(\varepsilon)$ describes the power spectral density of a time series as a function of frequency f, the variance of ε in the interval f_1, f_2 can be computed from

$$\text{var}(\varepsilon) = \int_{f_1}^{f_2} PSD(\varepsilon)\, df \tag{30}$$

and the power spectral density of the time derivatives $\dot{\varepsilon}$ and $\ddot{\varepsilon}$ is given by (e.g. Bower and Halpenny, 1987)

$$PSD(\dot{\varepsilon}) = (2\,\pi\,f)^2\; PSD(\varepsilon) \tag{31}$$

and

$$PSD(\ddot{\varepsilon}) = (2\,\pi\,f)^4\; PSD(\varepsilon), \tag{32}$$

which leads to a relative amplification of the high frequency components in differentiation. Equally valid is the conclusion that low frequency components are subject to a relative suppression in differentiation.

This effect of differentiation on two different spectral distributions of carrier phase errors is schematically shown in Fig.2. On the left, the carrier phase errors are assumed to be white noise yielding a constant PSD. Twofold differentiation leads to frequency rate errors proportional to the fourth power of frequency. On the right, the carrier phase errors are assumed to be dominated by low frequency components (red noise) with a PSD inversely proportional to the fourth power of frequency. In this case we obtain a flat PSD (white noise) for the frequency rate errors.

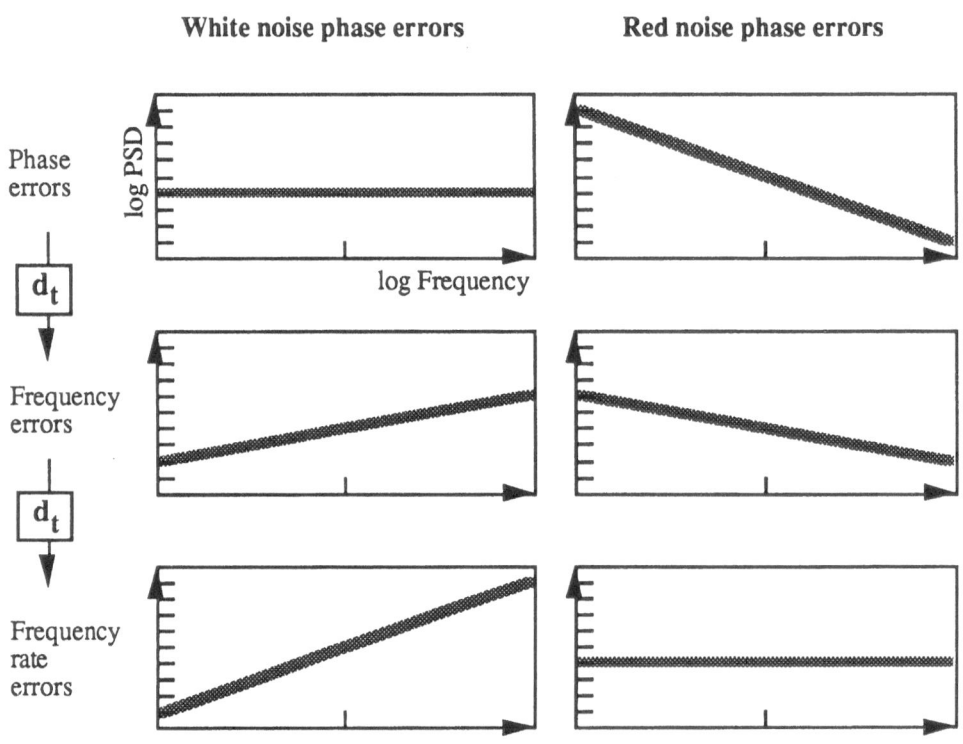

Figure 2: Error spectrum for phase, frequency, and frequency rate

The PSD distribution of the phase errors is important if the variance of the second derivative of the phase errors is of concern. This is the case in the application described in this paper: The second derivative of carrier phases is used to derive gravimeter platform accelerations (cf. section 4.1). Accordingly, the size of the errors in platform acceleration will depend on the error variance in second derivatives of phase. Let us assume that two different phase error time series with the same variance have PSD distributions as shown in the top plots of Fig. 2. Then the variance of the second derivative of the red noise will always be much smaller than the variance of the second derivative of the white noise, as the reader may easily verify from eqns. (30) through (32). Therefore, phase errors with "red" PSD distribution are preferable to white noise phase errors. As is shown below, in actual GPS phase measurements we see a mixture of both types of distribution.

4.3 Carrier Phase And Frequency Rate Accuracy

In this section we assess the accuracy and spectral error distribution of GPS carrier phase measurements and frequency rates obtained through numerical differentiation. The data analysed here were collected with dual frequency Texas Instruments 4100 GPS receivers on a very short static baseline. As shown by Georgiadou and Kleusberg (1988), linear combinations formed from this type of data can be used to directly analyse carrier phase errors. Simultaneous carrier phase measurements of four satellite signals were recorded every two seconds for a total duration of about 20 minutes. Fig. 3 shows the estimated PSD's for the four time series of carrier phase errors in the frequency range between 1 mHz and the Nyquist frequency of 0.25 Hz. Clearly recognisable in all four plots are two distinct regions. Above 0.02 Hz of frequency, the PSD's are fairly flat and represent band limited white noise. Below this frequency, the plots exhibit a f^{-3} to f^{-4} slope representing rapidly increasing spectral power density towards lower frequencies. It was shown by Georgiadou and Kleusberg (ibid.) that the flat level of about 20 mm^2/Hz PSD represents the receiver noise, and that the increase in PSD in the lower frequencies is primarily due to signal multipath and related phenomena.

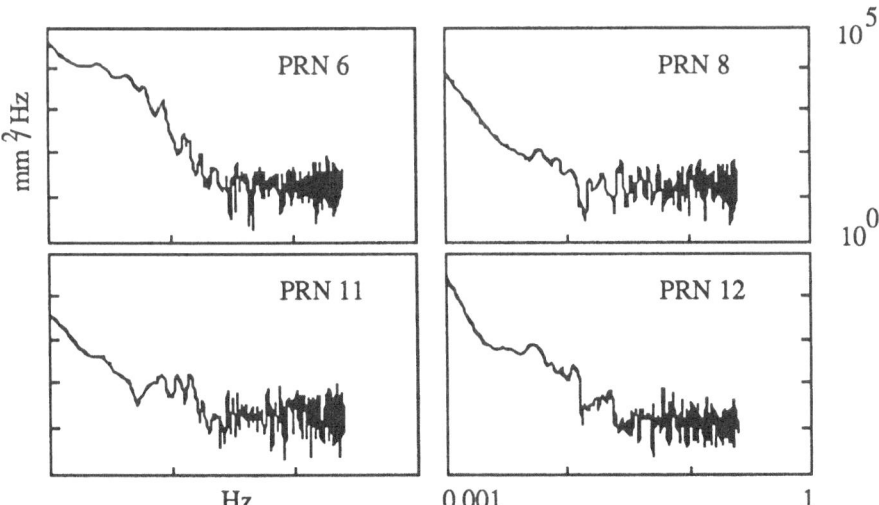

Figure 3: Carrier phase derived range error spectra

According to the discussion in section 4.2, differentiation amplifies the power density in the high frequency range and reduces the power density in the lower frequencies. The result of this process is shown in Fig. 4 depicting the PSD estimates for the second numerical derivative of the carrier phase observations (range accelerations). Again, two distinct regions are visible. Above a frequency of 0.02 Hz the PSD is proportional to f^4 , and results from the flat part in the power density of the carrier phase errors shown in Fig. 3. Below this frequency, the PSD's are flat at levels between 20 mGal^2/Hz and 100 mGal^2/Hz, resulting from the sloped part of the power density of the corresponding carrier phase errors in Fig. 3. The standard deviation sd corresponding to the PSD plots in Fig. 4 is given from eqn. (30) according to

$$ sd \ = \ \sqrt{\int_{f=0}^{0.25 \text{ Hz}} PSD \ df} \tag{33} $$

and amounts to about 80 mGal for the data analysed here. Instantaneous platform accelerations at about the same accuracy level can be computed from the second derivative of the carrier phases using procedures outlined in section 4.2. The resulting horizontal components of the gravimeter platform acceleration obviously satisfy the accuracy requirements as laid out in the discussion of eqns. (24) and (25).

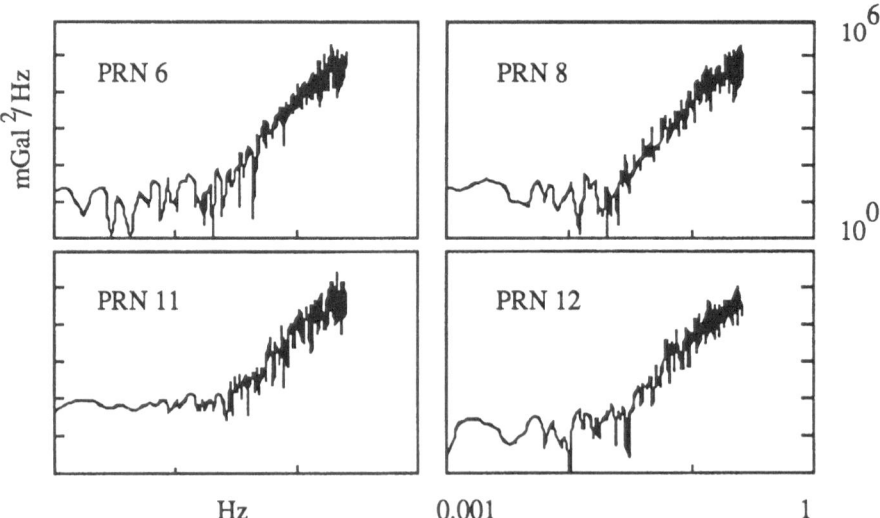

Figure 4: Frequency rate derived range acceleration error spectra

According to the specifications for airborne gravity surveys as listed in section 2., typical integration periods for an airborne gravity survey are of the order of 120 seconds. To correct these integrated gravimeter readings for platform accelerations, only average accelerations over 120 seconds are required. Therefore, all range acceleration (and range acceleration error) constituents with frequencies higher than 0.0083 Hz can be removed from the second derivative of the carrier phase observations. The error standard deviation of the remaining low frequency constituents can be computed from eqn. (30) and yields for a PSD level of 100 mGal^2/Hz

$$sd = \sqrt{\int_{f=0}^{0.0083 \text{ Hz}} PSD \, df} \approx 1 \text{ mGal}. \tag{34}$$

This means that the data analysed here allows for the computation of 120 second averages of satellite-to-receiver range accelerations from GPS carrier phase observations with an accuracy of about one mGal. These range accelerations (frequency rates) can be transformed into gravimeter platform accelerations at roughly the same accuracy level using procedures outlined in section 4.2. This accuracy level approaches the requirements for vertical platform accelerations as given in the description of eqn. (27).

5. CONCLUSIONS

It has been demonstrated in this paper how GPS measurements can be utilised for the separation of gravity from vertical accelerations caused by platform motion. GPS also can provide horizontal accelerations to be used for platform verticality control. The analysis of carrier phase measurements showed that GPS determined acceleration accuracy approaches the one mGal level, and therefore is sufficient for reconnaissance gravity surveys as described in section 2. The accuracy assessment was based on the analysis of GPS carrier phases collect with static receivers.

It can be seen in figures 3 and 4 that most of the acceleration errors are not caused by receiver noise but are attributable to signal multipath. If these signal variations can be avoided, the remaining acceleration errors will be smaller by one order of magnitude (cf. Fig. 3 and Fig. 4). For receivers actually moving, a slightly higher noise level can be expected. This will not change the conclusions regarding acceleration accuracies, since receiver noise plays a minor role only in the overall accuracy figure. As can be seen from the discussions in section 4.3, GPS determined acceleration accuracy increases with averaging time interval. Therefore higher accuracies can be expected for slow moving gravimeter platforms in balloons or airships.

ACKNOWLEDGEMENT

This research has been funded through a Natural Sciences and Engineering Research Council Strategic Grant entitled "Application of Differential GPS Positioning" and through a Research Agreement with the Department of Energy, Mines and Resources.

REFERENCES

Bell R, LaBreque J, Raymond C, Chayes D, Brozena J (1986) GPS vertical velocities: a source of vertical accelerations for airborne gravity surveys. AGU fall meeting, San Francisco, Ca.

Bomford G (1971) Geodesy. Oxford University Press, London

Bower DR, Halpenny JF (1987) On the technical feasibility of airborne gravity measurements. Internal report 87-3, Geophysics Division, GSC/EMR, Ottawa, Ontario

Brozena JM, Mader GL, Peters MF (1989) Interferometric Global Positioning System: Three-dimensional positioning source for airborne gravimetry. Journal of Geophysical Research 94 (B9): 12153-12162

Brozena JM, Eskinzes JG, Clamons JD (1986) Hardware design for a fixed-wing airborne gravity measurement system. Naval Research Laboratory Report 9000, Washington, D.C.

Cannon E (1987) Kinematic positioning using GPS pseudorange and carrier phase observations. Dept. Surveying Engineering Report 20019, The University of Calgary, Calgary, Alberta

Georgiadou Y, Kleusberg A (1988) On carrier signal multipath effects in relative GPS positioning. Manuscripta Geodaetica 13: 172-179

Kleusberg A (1986) GPS positioning techniques for moving platforms. Proc. ISPRS Symp. Stuttgart, FRG, ESA SP-252: 201-205

LaCoste LJB (1967) Measurement of gravity at sea and in the air. Reviews of Geophysics 5: 477-526

Lachapelle G, Lethaby J, Casey M (1984) Airborne single point and differential GPS navigation for hydrographic bathymetry. The Hydrographic Journal 34: 11-18

Landau H (1989) Precise kinematic GPS positioning. Part 1: Experiences on a land vehicle using TI4100 receivers and software. Bulletin Geodesique 63: 85-96

Mader GL (1986) Dynamic positioning using GPS carrier phase measurements. Manuscripta Geodaetica 11: 272-277

Otnes RK, Enochson L (1978) Applied time series analysis. John Wiley & Sons, New York Chichester Brisbane Toronto

Schwarz KP, Cannon ME, Wong RCV (1987) The use of GPS in exploration geophysics - a comparison of kinematic models. Pres. at General Assembly, IUGG, Vancouver, B.C.

Wells DE, Beck N, Delikaraoglou D, Krakiwsky EJ, Lachapelle G, Langley RB, Nakiboglu M, Schwarz KP, Tranquilla JM, Vanicek P (1986) Guide to GPS positioning. Canadian GPS Associates, Fredericton, N.B.

Precise Gravimetric Geoid Computations Over Large Regions

R. Forsberg
National Survey and Cadastre
Denmark

A.H.W. Kearsley
University of New South Wales
Australia

ABSTRACT:

Methods for gravimetric geoid computations are reviewed and compared on two long GPS profiles in Scandinavia and Australia. The comparison of GPS determined and gravimetric geoid undulations have yielded accuracies down to 12 cm r.m.s. in Scandinavia, and 22 cm r.m.s. in Australia, for solutions based on local gravity and height data (gravity only for Australia). In terms of accuracy of geoid height differences between neighbouring GPS stations (baseline lengths 50-100 km), the accuracy is typically 2 ppm, compared to 0.1 ppm for the whole 1500-2000 km long profile.

Comparisons are made between Stokes' integration by rings or grids, FFT methods and collocation, with the FFT and Stokes' approaches yielding similar results, implying that the planar approximation of the FFT method may be used for regions up to 2000 km extent without significant systematic errors. For Scandinavia additional comparisons are made regarding the use of detailed height data. Neglecting such data results in the geoid prediction error increases roughly by a factor of 2.

Comparison tests with various cap sizes in Stokes' integration indicate large systematic errors in the existing spherical harmonic geopotential models in Scandinavia. Tests are done with tayloring of improved spherical harmonic reference models, with good results in Scandinavia, but less satisfactory results in Australia, where the GPS profile used for control is located along the coast, with poor gravity data coverage offshore region.

1. INTRODUCTION

The geoid height provides, in a sense, a fourth dimension in satellite positioning, with the geoid height N required in order to convert ellipsoidal heights h to orthometric heights H

$$H = h - N \qquad (1)$$

With the widespread use of GPS, the precise determination of geoid heights has become a most important practical task of physical geodesy. Geoids computed by gravimetric means thus provide the capability for GPS to replace spirit levelling. On the other hand the use of precise GPS ellipsoidal heights along levelling lines provide an independent check of the capability of predicting geoids gravimetrically.

Numerous comparisons of GPS/levelling and gravimetrically derived geoid undulations have been published in recent years (e.g. Engelis et al. 1985, Schwarz et al. 1987, Kearsley 1988), and the capability to compute local geoids with error levels at a few cm has been demonstrated in several cases (see, e.g., Denker, 1988). In general most comparisons have taken place along GPS baselines, comparing differences in geoid height between two GPS points separated by distances typically in the range 10 to 100 km. Such results indicate typical GPS levelling accuracies around 2-3 ppm, expressed as a fraction of station separation.

However, such comparisons do not do the inherent accuracy of "GPS/gravimetric geoid levelling" full justice: Indeed the challenge presently is to determine the geoid consistently over large regions, so that a good match is obtained over extended GPS/levelling networks. Earlier results from Scandinavia have shown that a 2000 km GPS/levelling profile and a gravimetric geoid may fit within 15 cm r.m.s., or 0.1 ppm in terms of "baseline length" (Forsberg and Solheim, 1988).

At the longest wavelengths (the lowest spherical harmonics) the accuracy of global geopotential models imply a geoid accuracy of a few dm. Similarly on local scales (below 100 km) the geoid is rather easily recovered at the dm-level, provided sufficient gravity data is available. At longer wavelengths the available gravity data often tend to be less homogeneous, often based on different gravity surveys with the possibility for scale and datum problems, with data voids, and with systematic effects coming e.g. from inconsistent use of terrain corrections. These problems are reflected in the high-degree spherical harmonic reference fields available, where the higher harmonics (e.g. degree 60 to 180 or 360) may have large errors, and not properly represent the medium-wavelength field in many regions of the world.

In the sequel we will give examples of gravimetric geoid predictions by various methods, and give comparisons to extended GPS networks in Scandinavia and Australia, and with GEOSAT altimetry data off east Australia. We will especially consider the medium-wavelength errors, and the possibility of "tayloring" a spherical harmonical reference model to yield a better fit in a given region.

2. METHODS FOR GRAVIMETRIC GEOID COMPUTATIONS

Three operational geoid prediction methods will be considered here: Stokes' integration, least-squares collocation, and FFT methods. Common to all methods is that the effect of a high-degree spherical harmonic geopotential model

$$T_{ref} = \frac{GM}{R} \sum_{l=0}^{l_{max}} \sum_{m=0}^{l} (\Delta C_{lm} \cos m\lambda + \Delta S_{lm} \sin m\lambda) \bar{P}_{lm}(\sin\phi) \tag{2}$$

is used as a reference field, i.e. effects are subtracted from data prior to computations, and later added to the predicted geoid undulations.

2.1 Stokes' integration

Using Stokes' integral, geoid undulations are obtained as an integral over gravity anomalies, in principle extended over the whole surface σ of the earth

$$N = \frac{R}{4\pi\gamma} \int_{\sigma} \Delta g \, S(\psi) \, d\psi \tag{3}$$

Here R is the earth radius, γ a normal gravity value, and $S(\psi)$ the familiar Stokes' function. When a spherical harmonic reference field is used, $\Delta g - \Delta g_{ref}$ is used rather than Δg in (3). In this case the remote zone contributions in (3) will be small, and the integration is usually only extended to some spherical cap $\psi < \psi_{max}$. The selection of the "optimum" cap size can be quite tricky: If the used spherical harmonic reference field was perfect, then a cap size corresponding to the resolution of the spherical harmonic field should in principle be sufficient. However, as the reference fields may have quite large errors for the higher degrees, larger cap sizes are required in many cases. The use of increasing cap sizes do not always produce corresponding improvements in results, in some cases rather strange "oscillations" in prediction quality may be found (Kearsley, 1988).

In the numerical experiments described below two different software packages were used:

1) "RINT", direct ring integration of (3), where the spherical cap surrounding a prediction point is subdivided into concentric rings, and the integral is evaluated as a sum over the rings (Kearsley, 1986), with ring values being obtained by a weighted means interpolation.

2) Gridded Stokes integration (program "STOKES" of GRAVSOFT package) where a data grid is estimated by a fast quadrant-search collocation prediction algorithm ("GEOGRID"), followed by an integration of (3) over the grid compartments. To avoid the singularities of the integration close to the computation point (where $S(\psi)$ behaves like ψ^{-1}) a bicubical spline densification of the grid data is used in an inner zone.

2.2 Least-squares collocation

Least squares collocation requires the solution of a set of linear equations with dimension equal to the number of data

$$y_{pred} = \underline{C}_{xy} \ \underline{C}_{xx}^{-1} \ \underline{x} \qquad (4)$$

where $\underline{C}_{xx}$ and $\underline{C}_{xy}$ are the auto- and cross-covariance matrices between observations $\underline{x}$ and predictions $\underline{y}$ according to some selected analytical model approximating the empirical covariance function. The advantage of l.s.c. is the ability to utilize different data types, but the drawback is the need for the solution of a large set of linear equations, which in practice often prevents the utilization of all given data, and thus use of large "cap sizes". The collocation results given below for the Scandinavian GPS profile have thus been based on individual solutions in $3^\circ x \ 6^\circ$ blocks (Tscherning and Forsberg, 1986).

2.3 FFT methods

The FFT method is closely linked to the Stokes integration method, with the integral (3) effectively evaluated in the frequency domain rather than the space domain. The FFT method is based on the planar approximation, so it is essential to use a high-degree spherical harmonic reference field. In the usual formulation of the method the approximation $S(\psi) \approx const \cdot \psi^{-1}$ is used, yielding (Kearsley et al, 1985)

$$\tilde{N} \ = \ \frac{1}{\gamma k} \ \tilde{\Delta g} \tag{5}$$

with

$$\tilde{\Delta g}(k_x, k_y) \ = \ \iint\limits_{-\infty}^{\infty} \Delta g(x,y) \ e^{-i(k_x x + k_y y)} \ dxdy \tag{6}$$

$$k \ = \ (k_x^2 + k_y^2)^{1/2} \tag{7}$$

The approximation (5) implicitly assumes the gravity anomaly to be the vertical derivative of the anomalous potential. However, since for the spherical approximation

$$\Delta g \ = \ - \frac{\partial T}{\partial r} - \frac{2}{R} \ T, \qquad N \ = \ \frac{T}{\gamma} \tag{8}$$

an improved planar formula is obtained by

$$\tilde{N} \ = \ \frac{1}{\gamma(k - \frac{2}{R})} \ \tilde{\Delta g} \tag{9}$$

In this formula the "indirect effect" (the term 2T/R) is not neglected. Another important improvement of the planar approximation is the use of proper map projection for gridding the data, rather than using a geographic $\phi-\lambda$ grid, in order to avoid the errors caused by the meridian convergence at high latitudes. The FFT computations described below have

thus been done using a gridding/Fourier transformation program system
("GEOGRID/GEOFOUR") implementing full use of UTM grids.

3. USE OF AUXILARY DIGITAL TERRAIN MODELS

The use of terrain reductions of gravity data allow gravity anomalies to
be interpolated with higher accuracy in rugged topography, and atte-
nuates aliasing problems associated with insufficient sampling of gravity data caused by the high-frequency gravity field generated by the
topographic masses. When terrain reductions are used, they must be
applied systematically to both gravity and geoid. This may be done using
a "remove-restore" technique, where terrain effects are subtracted from
observation data and later added to the predictions, analogous to the
use of a spherical harmonic reference field. For details see Forsberg
and Tscherning (1981).

A convenient type of terrain reduction is the residual terrain model
(RTM) reduction, where only the high-frequency topographic irregulari-
ties are removed. In this case a smooth mean elevation surface must be
defined, and mountains above this surface is computationally "removed"
and valleys below "filled" with material of some assumed topography density. The advantage of the RTM reduction is that the effect on the geoid
is rather small, and that systematic errors in the digital terrain models used have less influence on long wavelengths, critical for precise
regional geoid prediction. The effects of the residual topography may be
expressed as

$$N_m(P) = G\rho \int\!\!\int_{-\infty}^{\infty} \int_{h_{ref}}^{h} \frac{1}{r} \, dzdxdy, \quad r = ((x-x_p)^2 + (y-y_p^2) + (z-z_p)^2)^{1/2} \qquad (10)$$

$$\Delta g_m(P) \approx 2\pi G\rho(h-h_{ref}) - c, \quad c = G\rho \int\!\!\int_{-\infty}^{\infty} \int_{h_p}^{h} \frac{z}{r^3} \, dzdxdy \qquad (11)$$

where G is the gravitational constant, ρ the density, and c the "classical" terrain correction. Both of the integrals in (10) and (11) may be
evaluated efficiently by FFT methods, for details see Forsberg (1985).

In the computations below, digital terrain models have been used only in
Scandinavia, because no dense DTM was available in Australia. In

Scandinavia the RTM reduction was performed relative to a 1° mean height surface, with gravity terrain effects computed from 1 km heights in the mountainous regions, and geoid effects computed from a 5' mean height grid. For more details see Forsberg and Tscherning (1986).

4. COMPARISON OF GRAVIMETRIC GEOID PREDICTIONS

Two long precise GPS/levelling profiles in Scandinavia and Australia have been used for evaluating the gravimetric geoid predictions. The Scandinavian profile used dual frequency receivers (TI4100), while the Australian profile used single frequency (Trimble) receivers, augmented with dual frequency observations over some lines.

The Scandinavian GPS profile (fig. 1), surveyed by Institut fur Erd-messung, Universitat Hannover, in cooperation with the national survey agencies, follows nearly 2000 km of first-order levelling lines, and ha-ve been tied to an absolute level (nominally WGS84) using geocentric coordinates of the Wettzell laser ranging station (Torge et al., 1988).

The Australian GPS profile (fig. 2) have been surveyed by the Public Works Department of N.S.W., and consists of a tide gauge network mainly along 1600 km of the south-east Australian coast (Macleod et al., 1988). The orthometric heights used for deriving geoid undulations at the GPS points by (1) have been based on tide gauge mean sea levels, but differs generally less than 10 cm from the levelled heights in the Australian Height Datum (Ibid., fig. 1). The absolute level of the GPS geoid heights has been set to agree with the geopotential model GPM2.

4.1 Scandinavian geoid predictions

In Scandinavia the available gravity data was initially screened using a 6' x 12' pixel search algorithm, leaving more than 20000 gravity points in the region. This thinned-out gravity data has subsequently been used in the various geoid prediction methods, either directly ("RINT" or "GEOCOL"), or using an intermediate grid covering 52°-73°N, 0°-36°E ("STOKES" or "GEOFOUR"). A data grid spacing of 6' x 12' (geographic) or 10 km x 10 km (UTM) was used in this case. The spherical harmonic model

GPM2 with l_{max} = 180 provided the best fit of data, and was used for all computations.

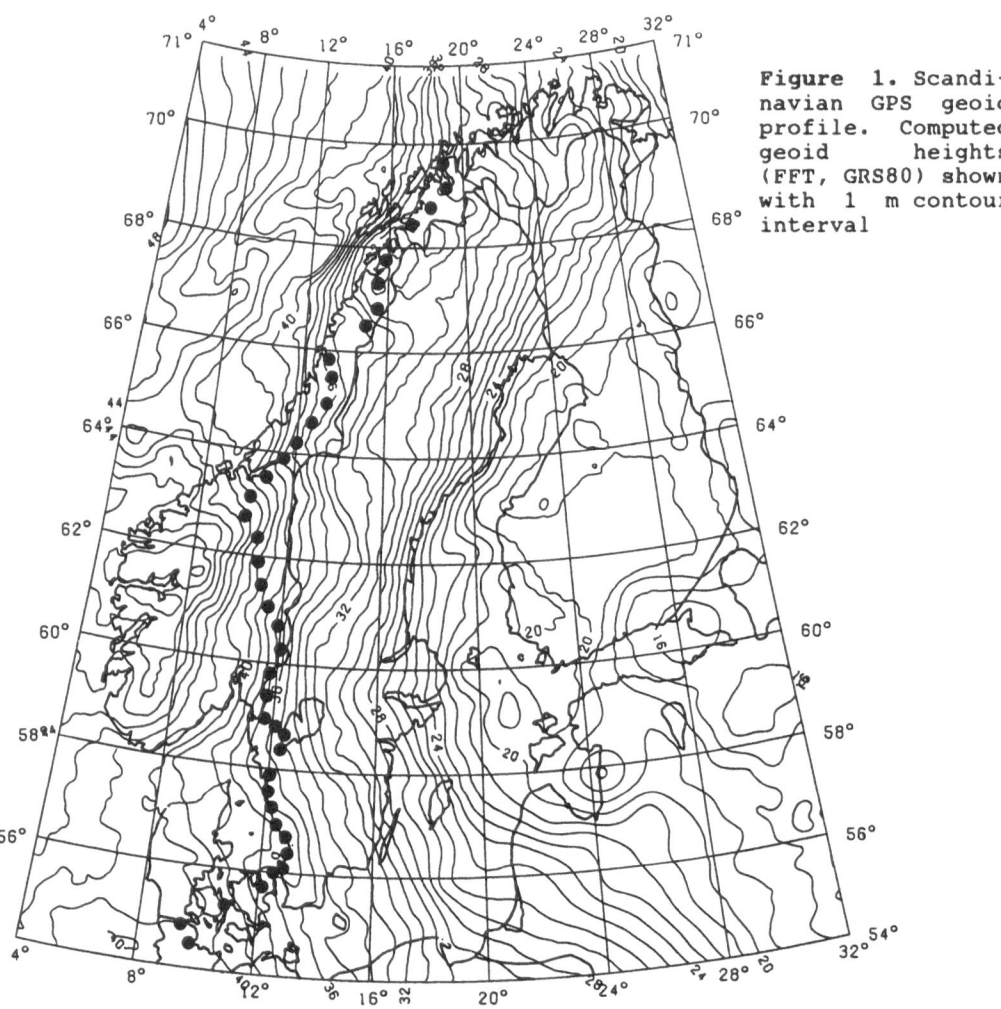

Figure 1. Scandinavian GPS geoid profile. Computed geoid heights (FFT, GRS80) shown with 1 m contour interval

Table 1 below gives the statistics of the original and reduced data (selected gravity anomalies and GPS geoid heights), and illustrates the smoothing obtained using GPM2 and the RTM reduction. In Table 2 the prediction results for various methods at the 41 GPS points are given.

Table 2 illustrates clearly that the used spherical harmonic reference field has long-wavelength errors in the region, and that gravity data subsequently must be taken into account in a large region (cap size) in order to obtain the best results. This also explains the poor performance of collocation, where computations have been done blockwise corresponding to an effective cap size of 1.5°.

The comparison of the two variants of the Stokes' integration ("RINT" or "gridded") shows that prediction results are dependent on the way the software handles data interpolation, and how biases in the data are treated (the small bias in the reduced gravity data evident from Table 1 are e.g. removed in the "gridded" variant, to make results comparable to FFT, but not done for "RINT").

The FFT results (12 cm r.m.s.) are surprisingly good, considering the approximations of the method. It must thus be concluded that even for a region as big as Scandinavia (20°) the planar approximation is still applicable for precise geoid predictions, provided proper x-y grids (UTM) and the better integral kernel (9) is used.

In terms of baselines (GPS geoid height differences over 50-100 km) the accuracies of the best solutions in Table 2 was 1.8 ppm. Over a 2000 km profile this ppm-value corresponds to 3.6 m, so it obvious that the use of "baseline accuracy" in ppm seriously underestimates the true potential for gravimetric geoid predictions and GPS to provide orthometric heights.

Table 1. Data statistics Scandinavia

	Δg (mgal)		N (m)	
	mean	std.dev.	mean	std.dev.
Original data	1.05	28.71	36.65	2.99
RTM/GPM2 reduced data	-1.13	14.63	0.25	0.68

Table 2. Prediction results: GPS minus gravimetric geoids

Prediction method and cap size ψ_{max} or variant of method	Differences (m)	
	mean	std.dev.
STOKES (gridded), cap size ∞	0.09	0.12
do , - 5°	-.19	.18
do , - 2°	.07	.17
do , - 1°	.13	.49
STOKES ("RINT"), cap size 4.5°	.85	.24
do , - 1°	.36	.52
COLLOCATION, $3^{\circ} \times 6^{\circ}$ solution blocks	-1.31	0.54
FFT ("GEOFOUR"), using UTM grid	0.06	0.12
do , geographical grid	-.12	.14
do , neglecting indirect effects by (5)	-.09	.18

Table 3. Stokes' prediction results with/without DTM's

Use of given terrain information	Differences (m)	
	mean	std.dev.
1 km x 1 km heights, normal RTM	0.09	0.12
No heights used at all (free-air Δg)	0.28	0.74
5' mean height grid used to interpolate mean free-air anomalies from gridded Bouguer anomalies without terrain corr.	-0.12	0.41
RTM reduction using 5' heights only, neglecting terrain correction c in (11)	0.23	0.21

The computations underlying Table 2 have all been done using RTM-reduction based on height grids down to 1 x 1 km resolution. In many regions, however, such detailed height data are not available, and it is therefore of interest to see how well predictions could be made without the use of dense height data (using no DTM at all or only 5' mean heights). Table 3 illustrates some results for different ways of using 5' mean heights, based on gridded Stokes' integration. From the table it can be seen that reasonable results (21 cm r.m.s.) can be obtained using even 5' height data only.

4.2 Geoid prediction in south-eastern Australia

In this case no terrain reduction was used due to lack of data, and a somewhat different data selection strategy was therefore adopted to minimize the influence of the topography in the grid prediction process (the region behind the N.S.W. coast is relatively rugged, with topographic heights up to 2000 m in the Snowy Mountains). The available free-air data from the Australian gravity data base was averaged in 6' x 6' cells, empty cells were subsequently predicted by the gridding routine. The FFT solution was part of a geoid solution covering eastern Australia and parts of the Coral and Tasman Seas (44°-10°S, 147°-160°E) in two blocks at a 6' grid. As reference field the model OSU86E, complete to degree and order 360, was used throughout.

Table 4 below shows data statistics for the 9500 averaged gravity data in the southern block (to 23°S), and the 39 GPS geoid values. Table 5 shows the results of various predictions. No collocation solution was attempted due to the computational work load involved.

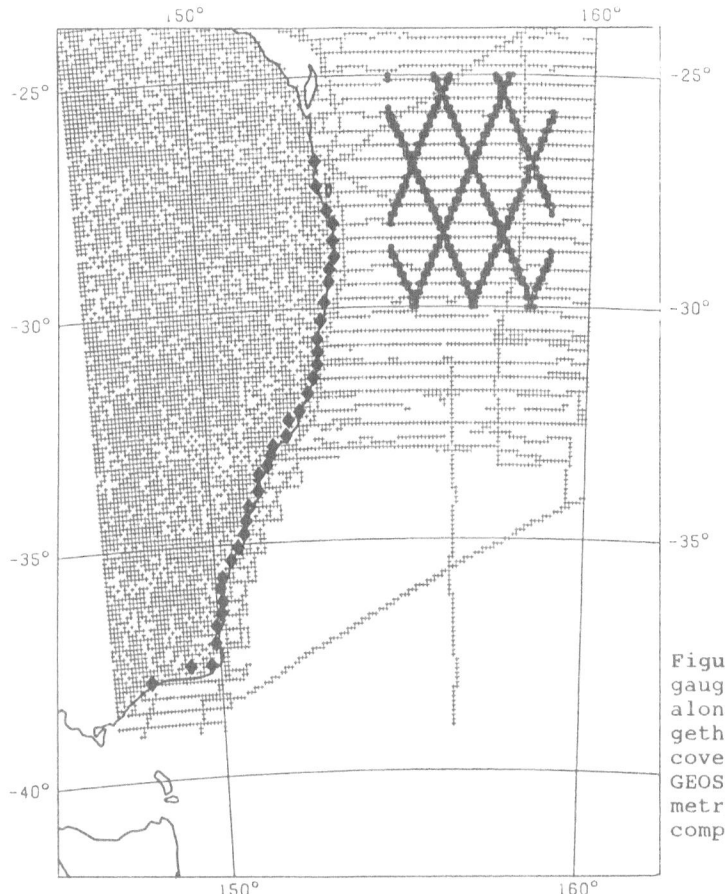

Figure 2. N.S.W. tide gauge GPS profile (dots along coast), shown together with gravity data coverage (ticks) and GEOSAT satellite altimetry data used for the comparison in sec. 4.3.

Table 4. Data statistics south-eastern Australia

	Δg (mgal)		N (m)	
	mean	std.dev.	mean	std.dev.
Original data	12.15	31.03	35.14	3.78
Data minus OSU86E	-3.25	19.24	-0.73	0.38

Table 5. Prediction results (GPS minus gravimetric geoid)

Geoid prediction method	Differences (m)	
	mean	std.dev.
Stokes ("gridded"), cap size 5°	-0.23	0.41
do , - 2°	-0.54	0.30
do , - 1°	-0.93	0.22
do , - 0.5^{o}	-1.00	0.22
Stokes ("RINT"), cap size 1.5^{o}	N/A	0.34
do , - 1°	N/A	0.27
do , - 0.5^{o}	N/A	0.20
FFT ("GEOFOUR")	-0.54	0.23

The results of Table 5 shows a degraded geoid prediction accuracy compared to Scandinavia, which to some degree probably reflects a poor gravity data coverage offshore, as well as the absence of topographic data. The ring integration software ("RINT") in this case provided the marginally better results, probably reflecting that this method uses the point gravity data directly.

In terms of long-wavelength errors coming from OSU86E, the integration with various cap sizes show these errors to be much smaller than in Scandinavia. In fact results seem to be worse when more gravity data than for the 0.5^{o} cap size are used, see also (Macleod et al., 1988, fig. 4). This illustrates the high quality of OSU86E in this region of the world, and that using a purely gravimetric geoid prediction method close to offshore areas with poor or absent gravity data may not yield such good results: here the collocation method with its ability to hand- le both gravity and satellite altimetry data would most likely yield better results.

4.3 Comparison of south-east Australian FFT geoid to GEOSAT data

The FFT solution of Table 5 covers a large part of the shelf off east Australia, so the same solution can also be evaluated offshore using sa- tellite altimetry data. In the present case GEOSAT data from 12 exact repeat missions, distributed in time through a year, have been selected in the region $35^{o}-25^{o}S$, $150^{o}-165^{o}E$, and subjected to a local cross-over adjustment (one orbit bias value per track). The adjusted data showed a cross-over variance of 25 cm, which to a large degree represented tempo- ral oceanographic signals associated with eddies of the East Australian current. By averaging in time these effects will be diminished.

The comparison of the averaged GEOSAT data to the FFT geoid is shown in Table 6 for the NW corner of the area (where gravity data exist at all), cf. Fig. 2. The results illustrate again problems caused by insufficient or too inaccurate offshore gravity data, as the FFT geoid solution (which compares to the GPS profile at 23 cm r.m.s.) apparently does not improve results at all. However, stationary sea-surface topography might also affect the results to some degree and make the comparison between the GEOSAT data and OSU86E too "optimistic".

Table 6. Comparison of the FFT solution to GEOSAT data.

GEOSAT mean sea-surface heights ($\approx$ N), m	mean	std.dev.
Original GEOSAT data (after adjustment)	35.14	3.78
Difference GEOSAT "geoid" minus FFT sol.	-2.35	0.46
- GEOSAT - - OSU86E	-2.43	0.41

5. TAYLORING OF SPHERICAL HARMONIC REFERENCE MODELS

The problems with errors in the spherical harmonic reference field, evident in Scandinavia, may be circumvented using "tayloring" of the expansion to local data (Weber and Zomorrodian, 1988). The basis of the method is to perform a spherical harmonic expansion of the residuals

$$\delta\Delta g = \Delta g - \Delta g_{ref} \qquad (12)$$

where Δg_{ref} is the gravity anomalies of an initial spherical harmonic expansion (1) with fully normalized coefficients ΔC_{lm} and ΔS_{lm}. The coefficients of the expansion of $\delta\Delta g$ are obtained by integration over a local area σ'

$$\left.\begin{matrix}\delta\Delta C_{lm}\\ \delta\Delta S_{lm}\end{matrix}\right\} = \frac{1}{4\pi}\iint_{\sigma'}\frac{R^2}{GM}\frac{1}{(1-1)\beta_1}\delta\Delta g\left\{\begin{matrix}\cos m\lambda\\ \sin m\lambda\end{matrix}\right\}\bar{P}_{lm}(\sin\phi)\,d\sigma' \qquad (13)$$

where β_1 are smoothing factors required if $\delta\Delta g$ is given as averages over blocks. When the integration (13) is carried out over a local area only, it is thus implicitly assumed that $\delta\Delta g$ is equal to zero outside σ'.

Modified potential coefficients are thus obtained by

$$\Delta C_{lm}' = \Delta C_{lm} + \delta\Delta C_{lm} , \quad \Delta S_{lm}' = \Delta S_{lm} + \delta\Delta S_{lm} \qquad (14)$$

These coefficients may then be used for providing new $\delta\Delta g$-values by (12), yielding new coefficients ($\Delta C_{lm}''$, $\Delta S_{lm}''$) and so on. In general, two to four iterations are required before sufficiently small $\delta\Delta g$-values are obtained.

The method was tested in both Scandinavia and Australia, using software originally developed by H.-G. Wenzel. In both cases the starting model was GPM2 to degree 180.

5.1 Geopotential model tayloring in Scandinavia

A set of 543 1°x 1° mean Δg values were derived from a 6'x 12' grid of reduced Δg-values, converted back to mean free-air anomalies using the mean height grids (the estimated grid point standard deviation from the collocation gridding algorithm was used to eliminate mean values being generated in areas without sufficient data).

The result of the iterative modification of GPM2 for degrees 10 to 180 is shown in Table 7, and the comparison of the new reference field to the GPS geoid data and point free-air gravity data is shown in Table 8. The fit to the GPS geoid data is also illustrated in Figure 3. It can be seen clearly that a significant improvement in the geoid has been ob- tained with the taylored model, the apparent slope error affecting GPM2 on the GPS profile being eliminated. Virtually no improvement is seen in the point gravity anomaly fit, which would also be expected, as the po- wer in the free-air gravity anomalies are dominantly coming from shorter wavelengths.

Table 7. Iterative improvement of taylored model

unit: mgal	mean	std.dev.
original 1°x 1° free-air data	-0.01	20.31
residuals $\delta\Delta g$ relative to GPM2	-.95	8.85
residuals $\delta\Delta g$ after 1. iteration	-.61	7.06
- 2. -	-.38	6.69
- 3. -	-.25	6.50
- 4. -	-.17	6.39

Table 8. Comparison of taylored GPM2 model to point values

| | GPS geoid data (m) | | free-air Δg (mgal) | |
	mean	std.dev.	mean	std.dev.
original data	36.65	2.99	1.05	28.71
data minus GPM2	0.27	0.72	-2.02	13.93
taylored model fit	0.41	0.37	-1.06	13.54

5.2 A taylored gravity field model for Australia

Similarly to Scandinavia a taylored GPM2 model has also been developed for the Australian continent. A total of 1086 1° x 1° mean free-air gravity anomalies have been averaged from the Australian gravity data base, selected within the region 44° - 10°S, 112° - 154°E where suffi-cient data existed.

The tayloring has been done with two iterations, again for degrees 10 to 180. The variance of the 1° x 1° gravity data changed from an original 23.0 mgal to 7.8 mgal when GPM2 was subtracted, and to 4.4 mgal in the taylored model. Fig. 4 shows the improved representation obtained in terms of 2° x 2° mean anomalies relative to GPM2 especially in the center of Australia.

The results of comparison to the tide gauge GPS profile, and a set of 0.5° x 0.5° mean gravity values are shown in Table 9. In this case a slight improvement is seen in the gravity data fit, analogous to Scandinavia, but a serious misfit has developed for the taylored model along the GPS profile. This misfit is mainly due to a slope error, and the probable conclusion is that lack of offshore gravity data has produ-ced serious "edge effects" in the integration (13), similar effects were also noted by Weber and Zomorrodian (1988). It therefore appears that the method of spherical harmonic tayloring should only be used with ca-re, the optimal solution to the problem is of course to submit improved average gravity values for inclusion in future global spherical harmonic solutions!

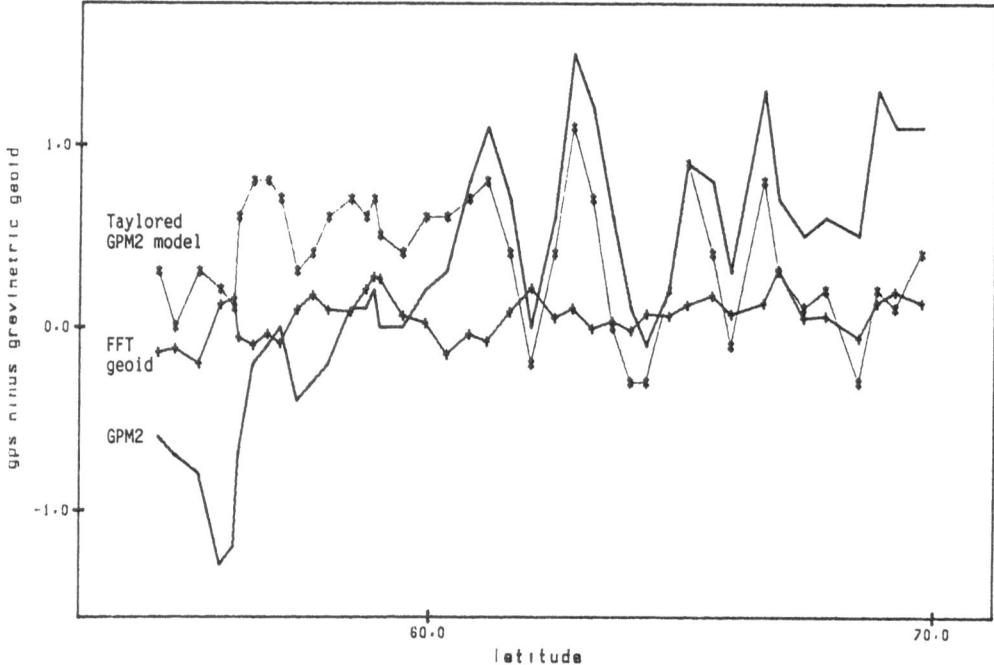

Figure 3. Differences GPS geoid minus GPM2 model (no points on curve), GPS minus taylored model ("*" on curve), and GPS minus FFT gravimetric geoid ("+"), from south to north in Scandinavia.

Table 9. Fit of taylored model in Australia.

	GPS profile (m) mean	std.dev.	0.5° $\overline{\Delta g}$ (mgal) mean	std.dev.
original data	35.14	3.78	0.37	25.79
- minus GPM2	(0.00)	0.34	-2.47	13.61
- minus taylored	(-.27)	1.09	-1.80	12.15

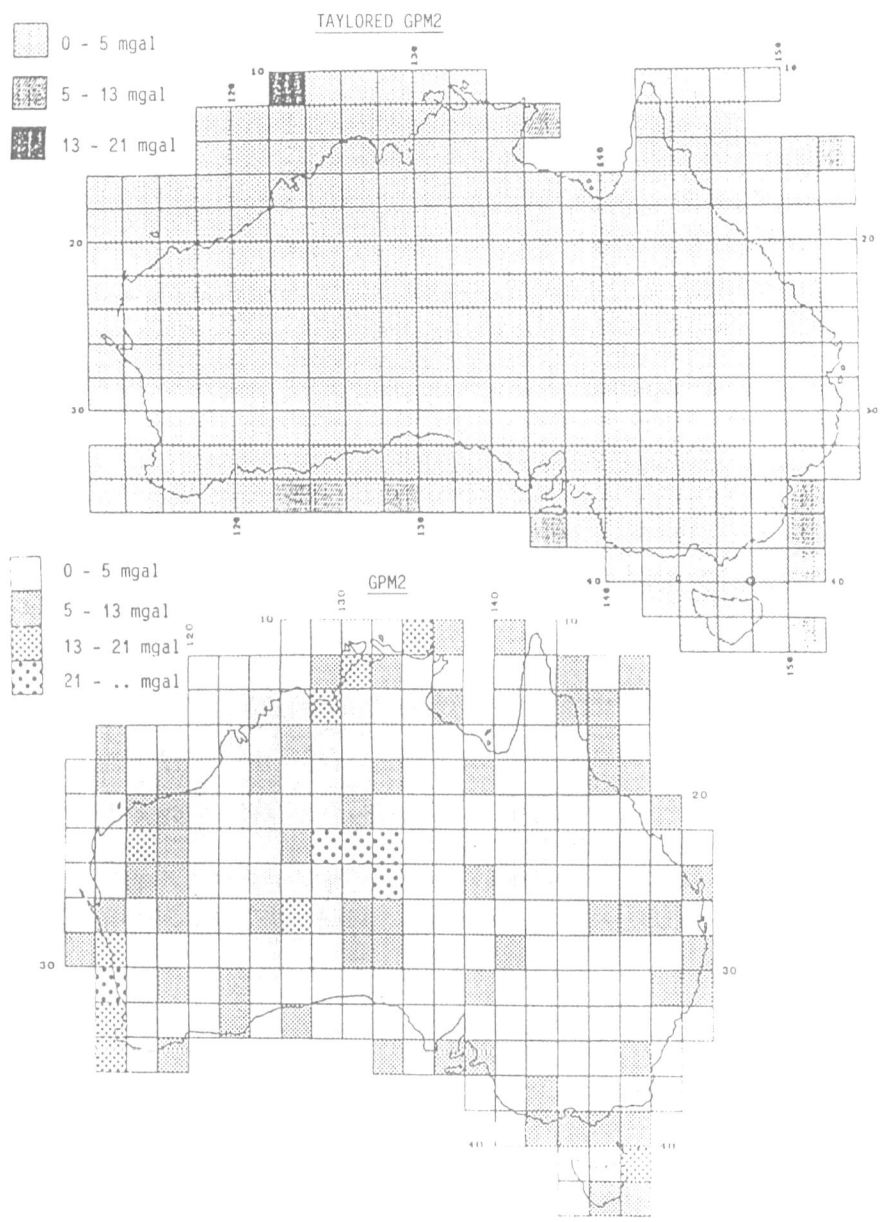

Figure 4. Magnitude of differences of 2° x 2° average free-air anomalies between observed values and mean anomalies of the new taylored model (upper), and GPM2 (lower) in Australia.

6. CONCLUSIONS

It has been demonstrated that gravimetric geoids may be computed with r.m.s. accuracies at the 10-20 cm level over baselines up to 2000 km, provided sufficient gravity and digital topography data is available. Such accuracy levels also demonstrate the high accuracy obtainable with GPS and levelling for providing "control" geoid undulations, for use e.g. in a national reference network if GPS methods are to replace levelling.

It appears that control of the long-wavelength errors of high-degree spherical harmonic reference fields play a major role in the quality of the long-range prediction results, and it is important to use prediction methods which are capable of taking into account gravity data for a large region in order to obtain the best results. Here the FFT method is especially attractive due to the computational efficiency, and the results for both Scandinavia and Australia show the method to yield comparable results to Stokes' integration, despite the various theoretical approximations underlying the method.

In areas where spherical harmonic models show a relatively poor fit, an alternative approach is spherical harmonic tayloring of an improved reference model, applicable to local gravity field modelling in the particular area only. Such tayloring should only be done with care: the estimation of good mean free-air anomalies can be quite difficult in mountainous areas, and the tayloring method can yield unreliable geoid results close to the edge of the data coverage, as seen in the Australian GPS profile.

ACKNOWLEDGEMENTS

Support for the first author's stay at the School of Surveying, University of New South Wales, has been provided by the Danish Natural Research Council and UNSW. Australian gravity data have been provided by BMR, Canberra, and GPS data by Department of Public Works, N.S.W. Scandinavian gravity and GPS data courtesy of the national survey agencies in the Nordic countries, Defense Mapping Agency, and Institut fur Erdmessung, University of Hannover, FRG.

REFERENCES

Denker, H. (1988): Hochauflosende regionale Schwerefeldbestimmung mit gravimetrischen und topographischen Daten. Wissenschaftlichen Arbeiten no. 156, Fachrichtung Vermessungsvesen der Universitat Hannover, 130 pp.

Engelis, T., R.H. Rapp, and C.C. Tscherning (1985): The precise computation of geoid undulation differences with comparison to the global positioning system. Geoph. Res. Lett. 11, no. 9, pp. 821-824.

Forsberg, R. and C.C. Tscherning (1981): The use of height data in gravity field approximation by collocation. Journ. Geoph. Res. 86, no. B9, pp. 7843-7854.

Forsberg, R. (1985): Gravity field terrain effect computations by FFT. Bull. Geod. 59, pp. 342-360.

Forsberg, R. and D. Solheim (1988): Performance of FFT methods in local gravity field modelling. Chapman conference on progress in the determination of the Earth's gravity field, Ft. Lauderdale, Florida, Sep. 1988, proceedings pp. 100-103.

Kearsley, A.H.W., M.G. Sideris, J. Krynski, R. Forsberg and K.P. Schwarz (1984): White Sands revisited: A comparison of techniques to predict deflections of the vertical. Rep. 30007, Dept. of Surveying Engineering, University of Calgary, Alberta.

Kearsley, A.H.W. (1986): The determination of precise geoid height differences using ring integration. Boll. Geod. Sci. Aff. 45, no. 2, pp. 151-174.

Kearsley, A.H.W. (1988): Tests on the recovery of precise geoid height differences from gravimetry. Journ. Geoph. Res. 93, B6, pp. 6559-6570.

MacLeod, R.T., A.H.W. Kearsley and C. Rizos (1988): G.P.S. surveys of mean sea level along the New South Wales coastline. Aust. J. Geod. Photogram. Surv. 49, pp. 39-53.

Schwarz, K.P., M.G. Sideris and R. Forsberg (1987): Orthometric heights without levelling. Journ. Surv. Eng. 113, no. 1, pp. 28-40.

Torge, W., T. Basic, H. Denker, J. Doliff, H.-G. Wenzel (1988): Long range geoid control through GPS techniques, final results. Preprint, Chapman conference on progress on the determination of Earth's gravity field, Ft. Lauderdale, Florida, sep. 1988.

Tscherning, C.C. and R. Forsberg (1986): Geoid prediction in the Nordic countries from gravity and height data. Boll. Geod. Sci. Aff. 46, pp. 21-43.

Weber, G. and H. Zomorrodian (1988): Regional geopotential model improvement for the Iranian geoid determination. Bull. Geod. 62, pp. 125-141.

A Revision of R. Mather's Work on the Determination of Stationary Sea Surface Topography and Global Vertical Datum Definition

B. Heck
University of Stuttgart
F.R.G.

ABSTRACT:

In the period 1973-1978 a great deal of R.S. Mather's work had been directed to practical approaches to the solution of the global vertical datum problem, aiming at an overall precision of about 15 cm. While in the earlier publications emphasis had been put upon purely terrestrial approaches (based upon the geodetic boundary value problem) R.S. Mather soon realized the potentialities of satellite altimetry. Some first numerical investigations based upon GEOS-3 altimeter data proved the principal applicability of this approach, but also indicated some deficiencies due to systematic errors in terrestrial gravity anomalies and insufficient precision of current geopotential models.

In the past decade the original ideas and concepts of R.S. Mather have been further developed. Nevertheless the final solution of the global vertical datum problem is not yet within sight. Besides the problems recognized by Mather and his colleagues, some additional error sources have recently been detected in both altimetry and terrestrial gravity data. The main deficiencies at present are due to long-period errors induced by the traditional mode of processing altimeter data, on the one hand, and by long-wavelength systematic errors in existing gravity anomaly data, on the other hand. Recent efforts aiming at the determination of quasi-stationary sea surface topography and global vertical datum definition are directed towards an improvement of the long-and medium-wavelength parts of geopotential models and at refinements in mathematical models of satellite altimetry.

1. INTRODUCTION

A great deal of R.S. Mather's work was devoted to the determination of the quasi-stationary sea surface topography considering applications in Physical Oceanography and Geodesy. The work in the period 1973 - 1978 was strongly influenced by the launch of the GEOS-3 satellite on April 9, 1975 and by the preparations for the launch of SEASAT-1 on June 26, 1978, both having been equipped with radar altimeters for measuring the distance between the satellite position and the instantaneous sea surface. After reducing the raw altimeter data for instrumental bias and periodic and short-wavelength effects in the instantaneous sea surface the resulting surface can be referred to a mean earth ellipsoid; the ellipsoidal height of the reduced sea surface is called quasi-stationary sea surface height. From theoretical considerations and experimental evaluations (based on oceanographic and geodetic levelling data) it has become clear in the sixties that the quasi-stationary sea surface cannot coincide with an equipotential surface of the earth's gravity field over the oceans. The difference in ellipsoidal heights of the two surfaces, the quasi-stationary sea surface topography, has amplitudes as great as 1 - 2 m and wavelengths as great as 8000 km. This phenomenon is primarily caused by oceanic currents, meterological effects and inhomogeneities of salinity and temperature distribution in the ocean water.

In Physical Oceanography the features of the quasi-stationary sea surface topography present a very important signal for evaluating the physical causes of this phenomenon. On the other hand, in Geodesy the same quasi-stationary sea surface topography is considered as an undesirable noise affecting high-precision geodetic data. The impact of quasi-stationary sea surface topography on gravity anomalies is twofold: First, gravity values measured by sea gravity meters can only be referred to the sea surface and not to the oceanic geoid, producing errors on the order of 0.3 - 0.6 mgal. Such a magnitude is often considered to be uncritical in view of the low precision of shipborne gravity meters; but it should be kept in mind that these model errors have a systematic nature, possessing wavelengths of some 1000 km. The second effect is related to the traditional way of fixing continental and regional levelling network datums using connections to tidal gauges at coastal sites. Due to the existence of sea surface topography the vertical datums of different zones are not consistent, i.e. there exists a regional bias in an arbitrary height system with respect to a unique global vertical datum. This bias theoretically should be constant over a vertical datum zone, amounting up to 1 - 2 m, the magnitude of the quasi-permanent sea surface topography. In practice the height datum bias must be considered as variable due to the behaviour of systematic levelling errors. Since the biased height enters the free-air reduction of surface gravity all free-air gravity anomalies over a specified vertical datum zone are systematically affected. These systematic errors have a strong impact on any quantity derived from gravity anomaly data, such as geoidal heights, height anomalies or deflections of the vertical.

This situation has formed the background of R.S. Mather's work concerning the determination of the quasi-stationary sea surface topography based upon surface gravimetry and satellite altimetry. Although it is principally possible to include oceanographic procedures for determining sea surface topography this approach has not been considered in R.S. Mather's work, probably in order to get independent results. Another type of additional data has also been left out of consideration, namely positional informations about surface points on land in the form of geocentric coordinates; it should be kept in mind that the present potentialities of positioning using the GPS could not be foreseen in the seventies.

In the following the basic ideas of R.S. Mather's work on the determination of stationary sea-surface topography and the impact of vertical datum inconsistencies are summarized. The deficiencies of the proposed approach for estimating the sea surface topography - already recognized by R.S. Mather - are discussed emphasizing the significance of orbital errors in the altimetry data and systematic errors in the gravity anomaly data. In addition an attempt is made to sketch the lines of further developments of R.S. Mather's original ideas during the past decade.

2. DETERMINATION OF QUASI-STATIONARY SEA SURFACE TOPOGRAPHY FROM GEODETIC DATA

The approach proposed by R.S. Mather for determining the quasi-stationary sea surface topography is based on the difference between the ellipsoidal heights of the sea surface and the geoid, both being independently determined. This is in contrast to oceanographic procedures aiming at a direct evaluation of sea surface topography (with respect to an isobaric surface) in one step. As a consequence the problem of determining the oceanic geoid with high precision forms a central part in R.S. Mather's publications. The term "geoid" is mostly used as a synonym for the quasi-geoid in the sense of Molodensky's theory, the difference being insignificant in oceanic areas. The evaluation of the (quasi-)geoid is treated within the framework of the geodetic boundary value problem forming the background of gravity field determination from terrestrial data.

2.1 Modelling Aspects and Data Requirements for the Solution of the Geodetic Boundary Value Problem

The key work concerning theoretical and practical aspects of the geodetic boundary value problem is contained in Mather (1973a). Many later papers, e.g. Mather (1973b, 1974a, 1974c, 1975a, 1978a) and Mather et al. (1976a) refer to the relationships and formulae derived in this early publication. In fact the formulation applied by R.S. Mather is very close to the scalar free geodetic boundary value problem as this is called recently (Heck 1989a, Sacerdote and Sansò 1986). In this concept gravity observations and geopotential numbers on the earth's surface are combined in order to determine the external gravity field of the earth and the ellipsoidal height of the surface points assuming the horizontal coordinates of the surface points to be given. The earth's surface acting as a boundary surface is defined as the boundary between the solid and fluid masses on one hand and the earth's atmosphere on the other hand. As a consequence the masses outside the boundary surface have to be considered separately, requiring reductions of the original observational data due to atmospheric and tidal effects of luni-solar origin. Besides periodical phenomena (which are generally not directly covered in the geodetic boundary value problem) stationary tidal effects have some consequences on the determination of the earth's gravity field. Aspects concerning the permanent earth tide have been discussed in Mather (1978c,d), while Anderson et al. (1975) deal with the influence of atmospheric mass attraction.

The theoretical approach applied by R.S. Mather to the solution of the geodetic boundary value problem is based upon the consistent use of Green's third identity. Another characteristic feature concerns linearization of the (originally non-linear) boundary conditions: instead of using a Somigliana-Pizzetti normal field - related to an equipotential ellipsoid - a 'higher' reference system in form of a geopotential model derived from satellite observations has been introduced by Mather (1974a). The original, rigorous but extremely complicated, relationships are simplified via "spherical", "planar" and other approximations (Moritz, 1980) leading to practically handable computation formulae. Aiming at a definition of height anomalies to ± 5 cm any approximation step has been justified by numerical estimates. As a result a closed, consistent theory of the geodetic boundary value problem to order $e^3 \approx 5 \cdot 10^{-4}$ (e = first numerical eccentricity of the earth ellipsoid) including computation formulae for practical applications has been worked out. Theoretical aspects discussed in Mather (1973a, 1974a) refer to the effects of topography and ellipticity, to first-degree terms caused by atmospheric masses, and to convergence of spherical harmonic representation of the gravitational potential. Other (smaller) effects are only mentioned but not considered in more detail: It should be worth mentioning that R.S. Mather draw attention to the effect of polar motion on terrestrial gravity observations as early as 1973; this phenomenon had to wait for a whole decade to be re-discovered by other authors (see e.g. Grafarend 1982, Heck 1984).

In the concept of the geodetic boundary value problem it is generally supposed that
the boundary data on the whole earth's surface refer to a unique horizontal, vertical
and gravity datum, and to the application of a rigorous free-air reduction formula. On
the other hand gravity anomaly data stored in data banks are mostly based upon regional
datums connected with national horizontal and vertical control networks; absolute gravity values are deduced from e.g. the IGSN-71, and free-air reduction is performed
using a globally constant vertical gradient of gravity $\partial g/\partial h$ = 0.3086 mgal/m. Due to
these and other sources of observational and meteorological origin long-wavelength
systematic errors are existent in present-day free-air anomaly data sets.

These deficiencies on the side of input data had been clearly discerned by R.S. Mather.
The order of magnitude of these systematic gravity anomaly errors is discussed in
Mather (1973a, 1974a, 1974c); a detailed evaluation of these effects for the area of
Australia has been reported in Mather et al. (1976b) and Mather (1978e). The main conclusion resulting from these numerical investigations states that the systematic errors
remaining in the corrected Australian gravity anomaly data set prevent a solution of
the geodetic boundary value problem with a precision better than ± 30 cm. Further reduction of systematic data errors should be feasible on the basis of additional absolute gravity determinations and threedimensional position fixes on the continents.
Similar conclusions have recently been obtained by Heck (1989b) for gravity anomaly
data sets covering other parts of the globe. Thus the present quality of data is still
insufficient for ± 10 cm sea surface topography determinations. A strong improvement
of this situation is expected from future satellite missions determining via satellite
gradiometry the long- and medium-wavelength parts of the gravity field one or two orders of magnitude better than terrestrial methods.

Determining sea surface topography is strongly connected with the definition of the
oceanic geoid forming the reference equipotential surface for describing the geometry
of the (quasi-stationary) sea surface. The problem of fixing a special equipotential
surface from the total set of level surfaces - the geoidal datum problem - has been
treated by Mather (1973b, 1974a, 1975b, 1978a, 1978b, 1978e) and Mather et al. (1978a,
1978b), taking into consideration geodetic as well as oceanographic criteria. The most
thorough and detailed discussion on this subject is contained in Mather (1978a) favouring the concept of the geoid as the level surface in relation to which the stationary
sea surface topography has zero mean on global sampling. It should be emphasized that
the term "geoid" is not meant to denote a level surface of the actual gravity field of
the earth inside the topographical or atmospheric masses; in contrast reference is made
to the atmospherically reduced, downward continued external field.

2.2 Determination of Quasi-Stationary Sea Surface Elevations from Satellite Altimetry

The second input data type required in R.S. Mather's approach into the determination of quasi-stationary sea surface topography is the geometry of the sea surface as derived from satellite altimetry. Assuming that the position and attitude of the satellite at the moment of a radar altimeter measurement is known from Laser tracking data the elevation of the instantaneous sea surface with respect to a conventionally adopted ellipsoid of revolution (arc length referred to the ellipsoidal normal) can be determined. The instantaneous sea surface height is composed of the quasi-stationary (permanent) sea surface height, the time variation of sea surface topography (meso-scale variability) and the ocean tidal signal (Mather 1976). The quasi-stationary sea surface results from reducing the raw altimetry data for instrumental bias and time-dependent effects using adequate reduction models.

The elevation of the instantaneous and quasi-stationary sea surface as deduced from satellite altimetry is subject to observational errors in the tracking and satellite altimeter data as well as to errors in orbit determination (Mather 1976, Mather et al. 1976a). While the observational errors can be expected to give rise to more or less randomly distributed effects in the elevation of the sea surface the orbit errors produce strongly systematic error patterns in the derived sea surface height. Orbit errors are mainly composed of two effects:

(i) Orbit distortion induced by errors in the coordinates of tracking stations held fixed in orbit determination.

(ii) Effects of errors of commission and omission in the gravity field model used in the orbit analysis.

The errors at (i) as well as the influence of other forces affecting satellite orbits are not critical with respect to precise orbit determination. In contrast the errors in gravity field models have proved to be the most problematic source of radial orbit errors; this fact has been corroborated by recent numerical investigations (see e.g. Engelis 1987, 1988, Schrama 1989). Radial orbit errors can partially be removed by minimizing the crossover differences of altimeter profiles; a predecessor of this technique has been applied to the analysis of GEOS-3 altimeter data by Mather et al. (1977). Local crossover minimization techniques based upon orbit error models described by constant bias and tilt parameters have extensively been used in the past decade for deriving regional and global sea surface models from GEOS-3 and SEASAT data (see e.g. Rapp 1979, 1982, 1985). Recent studies in satellite altimetry (Engelis 1987, 1988, Schrama 1989) proved that a major portion of long wavelength radial orbit errors cannot be detected from the local crossover minimization schemes as applied in the past. These rather simple approaches have to be replaced by a more analytical modelling of radial orbit errors, e.g. using global crossover minimization schemes or integrated approaches involving additional data sources. Other substantial improvements can be expected from the development of the next generation of geopotential models to be implemented in near future. In addition to the GEOS-3 and SEASAT data sets altimeter data from GEOSAT (launched in 1985) have recently become available; for the next decade ERS-1 is expected to be launched in 1990 and TOPEX/POSEIDON in 1991. As a consequence a huge amount of high precision and high resolution altimeter data will be available in the next decade. These prospects justify the assumption that the elevations of the quasi-stationary or permanent sea surface can be derived with a precision of ± 10 cm envisaged by R.S. Mather in the mid of the seventies.

2.3 Evaluation of the Quasi-Stationary Sea Surface Topography

Departures of sea level from an equipotential surface of the earth's gravity field, being called sea surface topography, have been suspected since a long time by oceanographers. These conjectures had been settled in the sixties and early seventies due to detailed evaluation of world-wide oceanographic data and geodetic levellings connecting coastal tide gauge stations. A compendium of informations including all available results from geodetic and steric levelling has been compiled in Mather (1974b). The quite puzzling fact that the sea surface slopes derived from geodetic and steric levelling did not fit together lead to a controversial discussion between oceanographers and geodesists in the seventies (see e.g. Fischer 1977), the disagreement being caused by the use of different reference surfaces. The principles of the oceanographic methods such as steric and dynamic (geostrophic) levelling have been discussed in Mather (1976, 1978b).

The concept proposed by R.S. Mather for evaluating the non-tidal sea surface topography is based on differencing between (tidal-free) sea surface heights and geoidal heights. Due to practical reasons the procedure is subdivided into three stages, assuming that the radial component of satellite orbital position is defined to at least ± 10 cm on a global basis (Mather 1978b):

(i) First the long wavelength components in the spectrum of quasi-stationary sea surface topography are determined, the shortest wavelength being equivalent to the degree of the geopotential model used in the course of orbital analysis. The corresponding formulae result from the theory of the geodetic boundary value problem. Including data based on regional levelling networks the mean sea level heights at all regional levelling datum zones can also be determined simultaneously, as long as the extension of any datum zone corresponds to the minimum wavelength.

(ii) Short wavelength contributions to the quasi-stationary sea surface topography are determined on a regional basis provided the surface gravity anomaly and satellite altimetry data show sufficient precision through the wavelengths of interest.

(iii) Time variations of sea surface topography are obtained from sea surface height variations taking care of geoid height changes with time.

This concept has been further discussed in the framework of ocean surface dynamics (Mather 1978b, 1978d) and has been applied to the determination of some dominant low-degree spherical harmonic coefficients of the stationary sea surface topography (Mather 1978e, Mather et al. 1978c). A summary of these ideas and some additional considerations can be found in Rizos (1980, 1982).

The procedure sketched above has been abandoned in recent years in favour of "integrated" approaches aiming at simultaneous improvements of 1) the geoid, 2) the radial position of the satellite, and 3) the quasi-stationary part of the sea surface topography. Such extended analysis models have become necessary due to the strong impact of gravity field model uncertainties acting upon parts 1) and 2), the geoid and the radial orbit error (see e.g. Engelis 1987, 1988, Schrama 1989). In contrast to R.S. Mather's assumption that the radial orbit position of the satellite can be defined to ± 10 cm, analysis of GEOS-3 and SEASAT altimetry data as well as simulation studies have shown that radial orbit errors as large as 2 metres may be induced by implementing the GEM9 gravity field model in orbit determination. The objective of an integrated approach - the improvement of 1) spherical harmonic coefficients of the gravitational potential and 2) coefficients of a surface spherical harmonic expansion of the quasi-stationary sea surface topography - can only be achieved by introducing additional observational data such as direct height measurements, PRARE or GPS positioning, or tracking data of satellites with different orbit characteristics. Another way out of this jungle may be expected from new gravity field mapping techniques such as satellite gradiometry. Although large efforts have been made in recent years for managing this extremely complicated problem a patent remedy is not yet in sight; much work has still to be done before the launch of the ERS-1 and TOPEX/POSEIDON satellites.

3. GLOBAL VERTICAL DATUM DEFINITION AND UNIFICATION OF GEODETIC LEVELLING DATUMS

Quasi-stationary sea surface topography has an important impact on the datum of regional or continental levelling networks, since "absolute" heights refer to the local mean sea level at one or several tide gauge stations situated near the coastline. Due to the existence of sea surface topography there exists no unique global vertical datum but many regional datums showing inconsistencies on the order of ± (1-2) m. There are many arguments for the unification of all regional geodetic levelling datums (Mather 1978a):

(i) Vertical datums on different land masses cannot be related to the same datum
     level surface (geoid) in the absence of direct levelling connections.

(ii) A comparison between geodetic levellings and oceanographic determinations of the
     sea surface slope is not possible on the basis of spatially scattered levelling
     networks.

(iii) Gravity anomaly data banks which are elevation dependent lack the resolution
     needed for high precision geodetic applications. Since gravity anomalies are
     related to the mean sea level at the datum gauge station and not to the geoid
     there exists an effect on combination solutions for geopotential models, espe-
     cially in higher harmonic constituents.

From a modern point of view other arguments for a unified global vertical datum defined to ± 10 cm can be added:

(iv) For monitoring sea level changes on various time and space scales on the back-
     ground of climate/ocean interaction a consistent reference frame is a basic pre-
     requisite.

(v) Future satellite altimetry and gradiometry missions require a global system of
     calibration points related to a globally consistent geodetic system.

For a discussion of modern aspects related to vertical control see e.g. Kaula (1987).

The impact of sea surface topography and vertical datum inconsistency on geodetic investigations had been considered already in the early publications by Mather (1973a,
1973b, 1974a, 1974b, 1975a, 1975b). The idea of a unification of regional geodetic
levelling networks on a global basis had first been formulated in Mather et al. (1976a).
In the sequel a complete procedure for levelling datum unification using satellite
altimetry has been worked out in Mather et al. (1978a, 1978b). Applications of this
approach to the estimation of the height of mean sea level at the Jervis Bay Datum
(Australia) and a Galveston Datum (North America) have been reported in Mather et al.
(1978a, 1978b) and Mather (1978e).

Estimates of the sea surface topography at datum gauge sites could be obtained directly from satellite altimetry according to the lines of reasoning presented in the previous section. Unfortunately such a direct approach is unsuitable due to abnormal reflections of the radar pulse so close to land areas and other difficulties arising
from switching between the ocean/land operation modes of the altimeter (Mather 1978a,
Mather et al. 1978b). As a consequence satellite altimetry may provide data of sufficient quality up to about 20 km from the coastline only. Having determined the heights
of the quasi-stationary sea surface topography in adjacent continental shelf areas -
based upon a globally defined geoid - these heights could be extrapolated in the
shallow continental shelf ocean to the coastal site. Extrapolation should be based on
the hydrodynamic equations resulting from the Lagrangean equations of motion (Mather

1976). The principal problem of this approach is the adequate modelling of frictional forces in continental shelf areas.

In view of these difficulties an alternate technique of determining the height of mean sea level at a coastal site has been proposed in Mather et al. (1978a, 1978b). This approach is essentially based upon the analysis of gravity anomaly data on regional vertical datums which cover areas greater than L^2 km^2; L denotes the shortest wavelength which can be resolved by the reference geopotential model with the desired precision of, say ± (5-10) cm. The height of mean sea level ζ_{sd} at the regional datum is obtained from forming observation equations of the gravity anomaly containing the quantity ζ_{sd} as an unknown systematic effect. Another systematic term occurring in the gravity anomaly observation equation is the zero-degree term W_o - U_o (W_o potential of the geoid, U_o potential of the reference level ellipsoid) which can be defined using satellite altimetry data in connection with a suitable definition of the oceanic geoid. This technique of determining ζ_{sd} relies heavily on gravity anomaly data banks; the only role played by satellite altimetry is in defining the bias term (W_o - U_o) selecting so-to-say an "absolute" global height reference surface.

Although practical applications of the approach to the determination of ζ_{sd} (Mather et al. 1978a, 1978b) have produced promising results some doubts have arosen concerning the unambiguous definition of an absolute oceanic geoid from satellite altimetry. The reason can be found in difficulties with respect to the calibration of the altimeter (Mather 1978e).

A variant on the approach sketched above has recently been formulated by Rummel and Teunissen (1987), avoiding some of the drawbacks of R.S. Mather's original method of determining ζ_{sd}. This technique is based on the solution of the linearized scalar free geodetic boundary value problem involving gravity observations on the whole earth's surface and levellings on the continents. Gravity anomalies which are affected by vertical datum inconsistencies are used as a basic data set. The unknown bias terms referring to different datum zones exert a direct as well as an indirect influence on the solution and can be estimated, e.g. in a least squares sense, if (in addition to gravity and levelling data) absolute highly precise geocentric positions of at least one site in any datum zone are given. This is a global approach, in principle, aiming at a simultaneous solution for ζ_{sd} for all datum zones. The complexity can be diminished restricting to only two datums, provided that a precise geopotential model becomes available purely from satellite methods.

In this context another procedure for vertical datum connections should be mentioned which has been investigated by Colombo (1980), Hajela (1983) and Rapp (1983). The idea behind this approach relies on accurate geocentric positions, terrestrial gravimetry, levellings, and low degree geopotential models. Since the theoretical background differs

to a large extent from R.S. Mather's approach it will not be discussed here in more
detail.

4. CONCLUSIONS

The considerations presented above have shown that R.S. Mather's work on the determination of quasi-stationary sea surface topography and global vertical datum definition
strongly influenced the developments in geodetic science and is still very up-to-date.
A large step forward can be expected for the next decade by the advent of satellite
altimetry missions such as ERS-1 and TOPEX/POSEIDON. With this high precision data at
hand the goals envisaged by R.S. Mather in the mid of the seventies might be reached.

ACKNOWLEDGEMENTS

Financial support from the German Research Foundation (Deutsche Forschungsgemeinschaft)
in the framework of a Heisenberg Research Grant is gratefully acknowledged. The author
is grateful to Prof. Dr. R. Rummel for stimulating discussions on this subject during
a two months visit at the Faculty of Geodesy, Delft University of Technology, Delft/
Netherlands in summer 1988.

REFERENCES

Anderson EG, Rizos C, Mather RS (1975) Atmospheric effects in Physical Geodesy. Unisurv
G 23: 23-41

Colombo O (1980) A world vertical network. Rep. No. 296, Dept. of Geodetic Science,
The Ohio State University

Engelis T (1987) Radial orbit error reduction and sea surface topography determination
using satellite altimetry. Rep. No. 377, Dept. of Geodetic Science and Surveying,
The Ohio State University

Engelis T (1988) On the simultaneous improvement of a satellite orbit and determination
of sea surface topography using altimeter data. manuscripta geodaetica 13: 180-
190

Fischer I (1977) Mean sea level and the marine geoid - an analysis of concepts. Marine
Geodesy 1: 37-59

Grafarend EW (1982) Six lectures on Geodesy and global Geodynamics. In: Moritz H,
Sünkel H (eds), Geodesy and Geodynamics, Mitt. Geod. Inst. TU Graz, Folge 41,
pp 531-685

Hajela DP (1983) Accuracy estimates of gravity potential differences between western
Europe and United States through Lageos satellite laser ranging network. Rep. No.
345, Dept. of Geodetic Science and Surveying, The Ohio State University

Heck B (1984) Zur Bestimmung vertikaler rezenter Erdkrustenbewegungen und zeitlicher Änderungen des Schwerefeldes aus wiederholten Schweremessungen und Nivellements. Deutsche Geodätische Kommission, C 302

Heck B (1989a) A contribution to the scalar free boundary value problem of Physical Geodesy. manuscripta geodaetica 14: 87-99

Heck B (1989b) An evaluation of some systematic error sources affecting terrestrial gravity anomalies. Bulletin Géodésique 63 (in print)

Kaula WM (1987) The need for vertical control. Surveying and Mapping 47: 57-64

Mather RS (1973a) A solution of the geodetic boundary value problem to order e^3. Goddard Space Flight Center, Greenbelt Md, Rep. X-592-73-11

Mather RS (1973b) The influence of stationary sea surface topography on geodetic considerations. Proc. Symposium Earth's Gravity Field, Univ. New South Wales, Sydney, pp 585-599

Mather RS (1974a) On the solution of the geodetic boundary value problem for the definition of sea surface topography. Geoph. J. Roy. Astron. Soc, 39: 87-109

Mather RS (1974b) Quasi-stationary sea surface topography and variations of mean sea level with time - an interim compendium. Unisurv G 21: 18-72

Mather RS (1974c) The gravity field and the definition of stationary sea surface topography. Publ. Zentralinst. Physik der Erde, Potsdam, No. 30 (2): 381-414

Mather RS (1975a) On the evaluation of stationary sea surface topography using geodetic techniques. Bulletin Géodésique 49: 65-82

Mather RS (1975b) Mean sea level and the definition of the geoid. Unisurv G 23: 68-79

Mather RS (1976) Some possibilities for recovering oceanographic information from the SEASAT missions. Unisurv G 24: 103-122

Mather RS (1978a) The role of the geoid in four-dimensional Geodesy. Marine Geodesy 1: 217-252

Mather RS (1978b) A geodetic basis for ocean dynamics. Boll. di Geodesia e Sc. Aff. XXXVII: 285-308

Mather RS (1978c) The influence of the permanent earth tide on determinations of quasi-stationary sea surface topography. Unisurv G 28: 76-83

Mather RS (1978d) The earth's gravity field and ocean dynamics. NASA Techn. Memo. 79540, Goddard Space Flight Center, Greenbelt Md

Mather RS (1978e) The geoid and continental gravity data banks: The role of satellite altimetry. Unisurv G 29: 1-9

Mather RS, Coleman R, Colombo OL (1976a) On the recovery of long wave features of sea surface topography from satellite altimetry. Unisurv G 24: 21-46

Mather RS, Rizos C, Hirsch B, Barlow BC (1976b) An Australian gravity data bank for sea surface topography determinations (AUSGAD 76). Unisurv G 25: 54-84

Mather RS, Coleman R, Rizos C, Hirsch B (1977) A preliminary analysis of GEOS-3 altimeter data in the Tasman and Coral Seas. Unisurv G 26: 27-46

Mather RS, Rizos C (1978a) On vertical datum definition from GEOS-3 altimetry. Proc. 2nd Int. Symp. Problems Related to the Redefinition of North American Geodetic Networks, Washington DC 1978: 589-597

Mather RS, Rizos C, Morrison, T (1978b) On the unification of geodetic levelling datums using satellite altimetry. NASA Techn. Memo 79533, Goddard Space Flight Center, Greenbelt Md

Mather RS, Lerch FJ, Rizos C, Masters EG, Hirsch B (1978c) Determination of some dominant parameters of the global dynamic sea surface topography from GEOS-3 altimetry. NASA Techn. Memo 79558, Goddard Space Flight Center, Greenbelt Md

Moritz H (1980) Advanced Physical Geodesy. H. Wichmann Verlag, Karlsruhe, Abacus Press Tunbridge Wells Kent

Rapp RH (1979) Global anomaly and undulation recovery using GEOS-3 altimeter data. Rep. No. 285, Dept. of Geodetic Science, The Ohio State University

Rapp RH (1982) A global atlas of sea surface heights based on the adjusted SEASAT altimeter data. Rep. No. 333, Dept. of Geodetic Science, The Ohio State University

Rapp RH (1983) The need and prospects for a world vertical datum. Proc. IAG Symposia, IUGG General Assembly, Hamburg, Vol. 2: 432-445

Rapp RH (1985) Detailed gravity anomalies and sea surface heights derived from GEOS-3/ SEASAT altimeter data. Rep. No. 365, Dept. of Geodetic Science and Surveying, The Ohio State University

Rizos C (1980) The role of the gravity field in sea surface topography studies. Unisurv S-17, School of Surveying, Univ. NSW Sydney

Rizos C (1982) The role of the geoid in high precision Geodesy and Oceanography. Deutsche Geodätische Kommission, A 96

Rummel R, Teunissen P (1987) Height datum definition, height datum connection and the role of the geodetic boundary value problem. Manuscript, Faculty of Geodesy, Delft University of Technology

Sacerdote F, Sansò F (1986) The scalar boundary value problem of Physical Geodesy. manuscripta geodaetica 11: 15-28

Schrama EJO (1989) The role of orbit errors in processing of satellite altimeter data. Netherlands Geodetic Commission Publ. on Geodesy, New Series No. 33

The Effects of a Possible Change in Climate on the Earth's Figure

K. Bretterbauer
Institute of Theoretical Geodesy
Technical University Vienna
Austria

ABSTRACT:

Recently, a possible change in climate caused by the greenhouse effect is much discussed by scientists but even more so in the public media. Generally it is maintained that the global mean temperature will rise by 3° - 5° C in the next few decades. As a consequence, melting of a considerable amount of ice and a world-wide rise of mean sea level by 1-3 m is assumed. These statements are too simple from a geodetic point of view. As the problem might be of vital importance for the next generations it is given a critical discussion.

1. INTRODUCTION

Presently an environmental problem obtains increasing attention by scientists but even more so by the public media, namely a possible change of climate caused by the greenhouse effect.

It is generally agreed by climatologists and environmental scientists that the output of wastes of modern civilisation produces a surplus of greenhouse gases as there are: carbon dioxide, halogen hydrocarbon, nitric oxide, and methane. These gases could be the cause of an increase in the global mean temperature by 3 - 5°C in the next few decades. The opinions on what then will happen are divergent. Two scenarios are offered:

a) A majority of experts maintains that the increasing temperature will melt part of the polar ice sheets (and the mountain glaciers) causing mean sea level to rise *uniformly* by 1 - 3 m. Subsequently, low-level countries, like the Netherlands will become flooded. Unfortunately, these dismal prospects are willingly taken up and commented on by the public media.

b) The other scenario states that rising global temperature will bring about increased evaporation and increased precipitation which in Antarctica and Greenland will fall in the form of snow as the surface temperatures there are far below freezing point. Hence, the great ice sheets will *increase* for a considerable time, making mean sea level fall. Finally, the increased precipitation should have a cooling effect, and some kind of a self-regulating mechanism will tend to stabilize the climate [Marcinek, 1977; Schwarzbach, 1974].

Of course, it is not up to a geodesist to judge which scenario is correct, but some comments may be in order. As to the first scenario, the expression "melting of the polar ice sheets" often is misinterpreted as people tend to include the floating ice masses of the Arctic Sea. If these masses should melt, nothing at all will happen to mean sea level, but would contribute quite a lot to evaporation thus favouring the second scenario.

Actually, the interaction of greenhouse gases, climate, and ice sheets is not yet fully understood. It can't really be imagined that sea level could rise by some meters in only a few decades. Anyhow, the problem seems to be connected with the transport of large masses. Hence, the problem gains a geodetic aspect as it affects the figure and the gravity field of the Earth. If those experts in climatology and environmental sciences are right, the problem could be of utmost importance already for the next generation. It therefore is somewhat surprising that the apparently pressing problem has not yet been taken up by the IUGG by establishing an interdisciplinary study group.

2. THE GEODETIC PROBLEM

As we know very little about the future evolution of our planet, the following computations are fictitious and intended only to give an outline of the geodetic aspects of the problem. Accordingly, the investigation is based on the following assumptions:

a) In agreement with the second scenario it is assumed that by evaporation a water layer of 1.5 m will be extracted from the oceans and deposited as a uniform ice layer on Greenland and Antarctica, respectively. The ocean area is $360 \cdot 10^6$ km^2, the total ice area of Greenland and Antarctica about $15.3 \cdot 10^6$ km^2. Hence, the additional ice layer would grow to a thickness of about 39 m (= 35.2 m water equivalent).

b) The computations are based on a rigid Earth. This is justified by the fact that the masses transported from the oceans to the ice sheets, large as they are, still are small as compared to the masses of the continental crust. Loading effects therefore are ignored. Furthermore, the process is assumed to take place in the rather short time of a few decades. For that reason isostatic rebound also is neglected.

c) The effect of the mass transport on gravity, on the elements of the inertia tensor, and on polar motion, though of great importance, are beyond the scope of this paper, and will be treated in a subsequent investigation. The study also is restricted to the change of mean sea level. Hence, the variation of the equipotential surface on the continents is *not* treated here.

3. THE SOLUTION OF THE PROBLEM

Similar problems have repeatedly been treated in the literature [Kivioja, 1967; Farrell et al., 1976; Nakiboglu et al., 1981; Stolz, 1976; Bretterbauer, 1982]. In view of the rather vague suppositions, a very simple procedure was chosen. All areas where variations of masses occur, i.e. the oceans, Greenland, and Antarctica, were divided into a total of 507 compartments of 10° x 10° size in moderate latitudes, and of up to 40° x 10° near the Poles. Thus, Greenland was substituted by 7 compartments, Antarctica by 41 compartments. The computations then are performed in three steps.

3.1 Deformation of the Potential Surface

In the first step, the masses to be transported were considered (positive or negative) immaterial condensed surface layers. If mass is added to or removed from a compartment, the level surface of potential V becomes displaced. Consider a spherical Earth of radius R, and of mass M. Then its potential is:

$$V = \frac{G.M}{R}, \qquad (G = \text{gravitational constant}). \qquad (1)$$

If a (point-) mass m is placed on the level surface V at P (cf. Figure 1) the surface of equal potential will be shifted from Q to Q′. The corresponding expression for V then is:

$$V = G\left(\frac{M}{R + h} + \frac{m}{d}\right), \qquad (2)$$

with d being the distance PQ, and h the vertical displacement of the equipotential surface. Equating equations (1) and (2) and considering that h is very small as compared to R, gives:

$$h = \frac{m}{M} \cdot \frac{R^2}{d}. \qquad (3)$$

where

$$d = 2 \cdot R \cdot \sin(\psi/2). \qquad (3a)$$

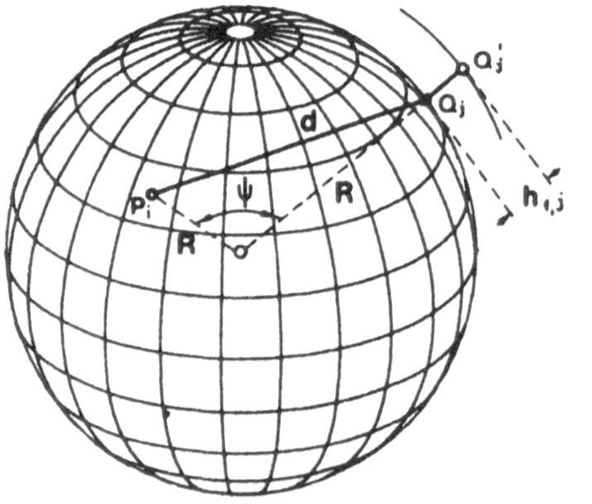

Figure 1

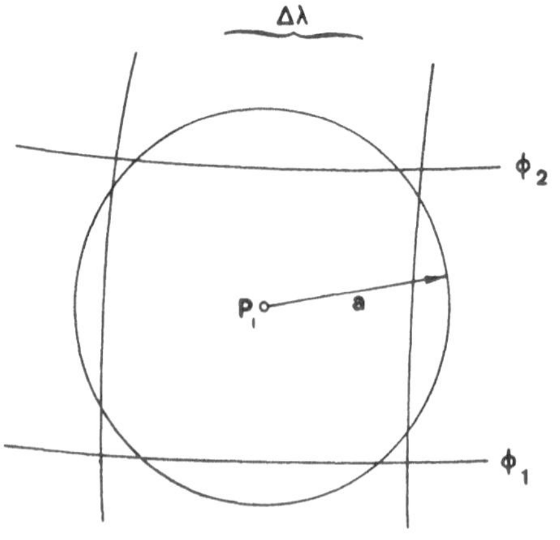

Figure 2

The central angle ψ (Figure 1) follows from the cosine-law of spherical trigonometry. For the numerical computation equations (3) and (3a) may be combined to

$$h_1 = 5.31 \cdot 10^{-19} \cdot m / \sin (\psi / 2) . \qquad (3b)$$

with h_1 in meters and m in kg.

Hence, for a given mass m the product $h \cdot d$ = const., i.e. h has a maximum where d is minimal. Formula (3b) allows to compute the displacement of the level surface for all 506 compartments except that one where the mass m is situated. Equation (3) does not apply for d approaching zero. As it is rather difficult, but also not necessary, to calculate the potential of a spherical layer bounded by meridians and parallels, these compartments were substituted by thin flat circular disks of equal area (cf. Figure 2). Let R be the radius of the Earth, φ the latitude, and $\Delta\lambda$ the difference in longitude, then the radius of the disk of equal area is:

$$a = R \cdot \sqrt{(\sin\varphi_2 - \sin\varphi_1) \cdot \Delta\lambda / \pi} , \qquad (4)$$

and its mass:

$$m = a^2 \pi \cdot t \quad \text{(with density = 1),} \qquad (4a)$$

if t means the thickness of the layer (i.e. - 1.5 m for the oceans, + 35.2 m for Greenland and Antarctica).

The potential of this thin circular disk is:

$$\delta V = 2\pi G \cdot a \cdot t , \quad \text{(density = 1).} \qquad (5)$$

According to Bruns' formula, this causes a displacement of the level surface in P by

$$h_2 = \frac{\delta V}{\gamma} , \qquad (6)$$

with γ being the mean gravity value of $9.80 \ m \cdot s^{-2}$. Formulas (5) and (6) can be combined for the practical computation to give:

$$h_{2,i} = 4.265 \cdot 10^{-5} \cdot a_i \cdot t_i , \qquad i = 1 \ldots \ldots 507; \qquad (6a)$$

with $h_{2,i}$ and t_i in meters, and a_i in km.

3.2 Shift of Geocenter

If a water layer is removed from the oceans and deposited in the form of ice on Greenland and Antarctica, mass is transported from mid-latidudes and deposited primarily near the South Pole. This results in a shift of the center of gravity. This shift of the geocenter needs some special consideration. If, as usual, the effect of a disturbing mass distribution is expanded into spherical harmonics, there are no terms of first order representing the coordinates of the geocenter. The reason is that disturbed as well as undisturbed mass configurations are by definition assumed to have the same center of gravity. In the present problem, however, conditions are different. The mass configuration before transportation of water takes place may represent the undisturbed equipotential, i.e. mean sea level. The removal of a water layer and deposition somewhere else now is a real transportation of masses which does not occur when disturbing masses are used to compute a geoid with respect to its level spheroid. Both surfaces, geoid and level spheroid, have identical centers of gravity. In the present problem the undisturbed former geocenter still is represented by the continents. And exactly the variation of sea level with respect to this old geocenter is wanted.

Again it suffices to treat all 507 compartments as point-masses. The components of the displacement vector caused by a single compartment are:

$$\delta\xi_i = \frac{m_i}{M} \cdot R \cdot \cos\varphi_i \cdot \cos\lambda_i$$

$$\delta\eta_i = \frac{m_i}{M} \cdot R \cdot \cos\varphi_i \cdot \sin\lambda_i \qquad\qquad [7]$$

$$\delta\zeta_i = \frac{m_i}{M} \cdot R \cdot \sin\varphi_i ; \qquad\qquad i = 1 \ldots\ldots 507.$$

Summing up all individual vectors gives the vector of the total shift of the geocenter:

$$s = \begin{bmatrix} \xi \\ \eta \\ \zeta \end{bmatrix} = \begin{bmatrix} 0.070 \text{ m} \\ 0.075 \text{ m} \\ -0.361 \text{ m} \end{bmatrix} . \qquad\qquad [8]$$

The absolute value of the displacement is then:

$$s = 0.362 \text{ m} \qquad\qquad [8a]$$

pointing in the direction:

$$\varphi_o = -74°\!.2 ; \qquad \lambda_o = 47°\!.1 \text{ East of Greenwich.} \qquad\qquad [8b]$$

In that connection it is interesting to note that the values (8b) are *independent* of the amount of masses transported but depend only on the present configuration of continents and oceans, and on the geographical position of Greenland and Antarctica. It may further be pointed out that the secular wandering of the Mean Pole tends roughly in the direction of (8b), (Sigl, 1975).

From vector (8) now the component perpendicular to each compartment has to be calculated. With Θ being the angle between vector **s** and the normal vector of the compartment the effect of the shift of the geocenter on the compartment is:

$$h_{3,i} = s \cdot \cos \Theta_i , \quad \text{with} \tag{9}$$

$$\cos \Theta_i = \sin \varphi_i \cdot \sin \varphi_o + \cos \varphi_i \cdot \cos \varphi_o \cdot \cos (\lambda_i - \lambda_o). \tag{9a}$$

3.3 Equality of Water Volume

Now the total height variation of the level surface can be derived for each compartment:

$$h_i = \sum_{j=1}^{507} \delta_{ij} \cdot h_{1,i} + h_{2,i} + h_{3,i} , \quad \delta_{ij} = \begin{cases} 1 & \text{if } i \neq j \\ 0 & \text{if } i = j \end{cases} . \tag{10}$$

Up to now the disturbing masses have been treated as condensated immaterial surface spreads. The redistribution of attracting masses in form of condensated matter only changes the shape of the ocean surface but not its volume, as the amount of ocean water has not yet been altered. That means that the summation over all oceanic compartments

$$\sum^{\text{oceans}} (h_i \cdot \text{Area}_i)$$

should vanish. If this is not the case all oceanic h_i have to be reduced by the quantity:

$$\delta h = \frac{\sum^{\text{oceans}} (h_i \cdot \text{Area}_i)}{\text{ocean area}} . \tag{11}$$

This check for equality of ocean volume resulted in:

$$\delta h = - \ 0.032 \ \text{m} . \tag{12}$$

Finally a real water layer of 1.5 m has actually to be removed from the oceanic compartments. Thus, the final formula for calculating the change of mean sea level reads:

$$h_i = \sum_{j=1}^{507} \delta_{ij} \cdot h_{1,j} + h_{2,i} + h_{3,i} + 0.032 - 1.50, \quad \delta_{ij} = \begin{cases} 1 & \text{if } i \neq j \\ 0 & \text{if } i = j \end{cases}, \tag{13}$$

all terms in meters.

4. THE RESULT

First, the change of mean sea level for some prominent coastal zones are given: North Pole: -1.88 m; Thule (Greenland): -1.76 m; Hudson Bay: -1.91 m; Vancouver: -2.01 m; New York: -1.90 m; The Netherlands: -1.82 m; Mediterranean Sea: -1.73 m; Hawai: -1.93 m; Los Angeles: -1.99 m; Tokio: -1.93 m; Rio de Janeiro: -1.40 m; Cape Horn: -1.03 m; Cape of Good Hope: -1.12 m; Perth (Australian West Coast): -1.16 m; Sydney Harbour: -1.29 m.

The change of sea level at the coast of Antarctica varies between -0.33 m and and +0.35 m. Because of the attraction of the additional ice layer sea level at the Ross Ice Shelf even rises by +0.50 m despite the fact that a water layer of 1.5 m was skimmed off the oceans. Figures 3 and 4 show maps with lines of equal sea level change (contour interval 0.25 m).

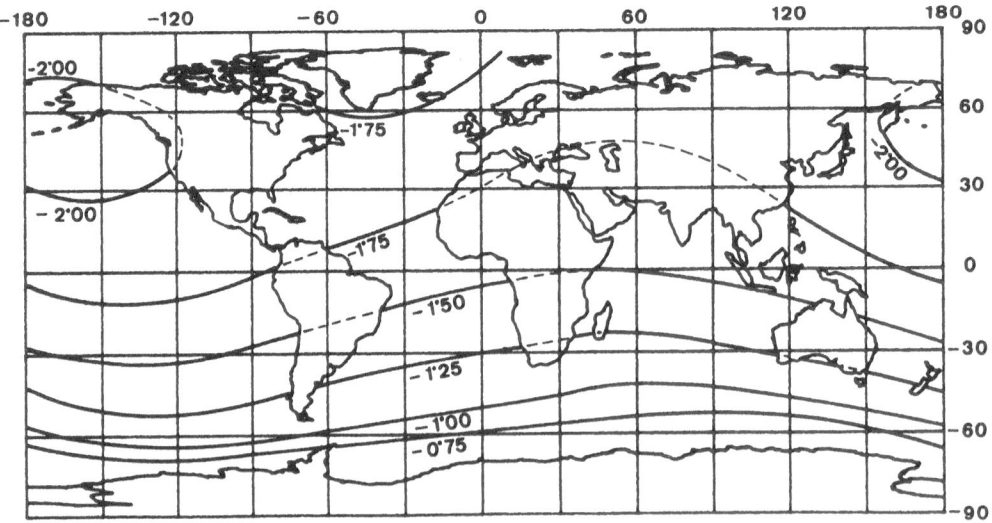

Figure 3

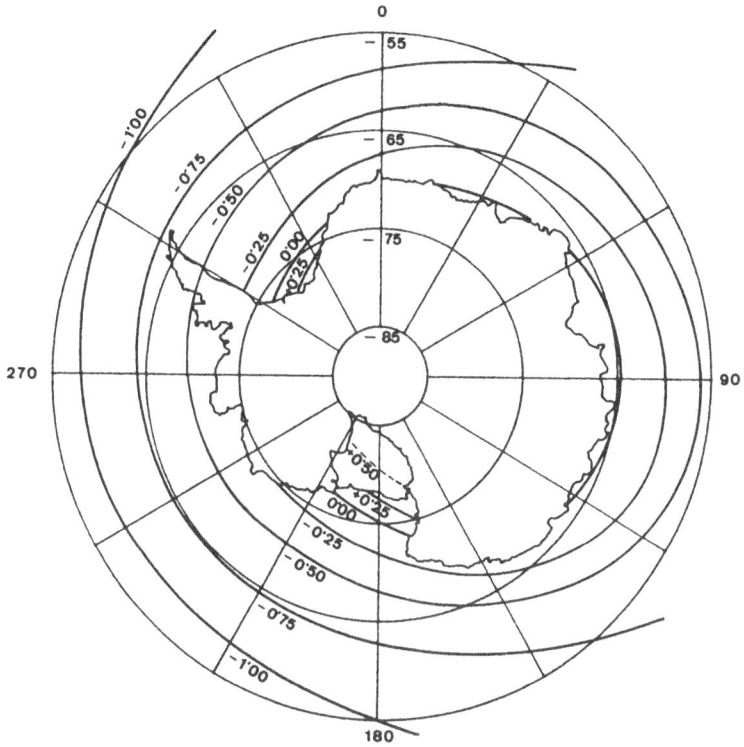

Figure 4

The redistribution of masses also changes the principal moments of inertia A, B, and C in units of 10^{30} kg·m^2 by:

$$\delta A = +\ 6.17$$
$$\delta B = +\ 7.44 \qquad\qquad [\ 14\ ]$$
$$\delta C = -\ 14.35\ .$$

The moment of inertia C about the rotational axis diminishes because mass was transported from low and moderate latitudes towards the Poles, i.e. closer to the rotational axis. If ω stands for the angular velocity of the Earth's spin, and T for its rotational period of 86164^s, the law of conservation of the rotational momentum reads:

$$C \cdot \omega = 2\pi \cdot \frac{C}{T} = \text{const.} \qquad\qquad [\ 15\ ]$$

If C diminishes, T also must diminish by:

$$\delta T = \frac{\delta C}{C} \cdot T\ . \qquad\qquad [\ 15a\ ]$$

Taking the known value $C = 8.036 \cdot 10^{37}$ kg· m^2 and for δC from (14), the length of day will be shortened by

$$\delta T = - 0.015^s .$$
(16)

The shift of the geocenter and the variation of the moments of inertia A and B would excite an additional polar motion. This effect will be treated in a separate study.

5. CONCLUSIONS

A change of climate in the next few decades as predicted by climatologists and other experts would have a considerable effect on the geometrical and physical parameters of the Earth, above all on mean sea level. This might turn out to become a vital problem for low-level countries and for the life of future generations. As we do not really comprehend how the great ice sheets in Greenland and Antarctica will react to increasing global temperatures, it is of utmost importance to monitor the ice budgets. Hence, supervision of the behaviour of global mean sea level becomes the most important task for geodesy in the future. As mean sea level depends on a variety of causes this should be done by observation of tide gauges, of the length of day, of absolute gravity at various selected sites, and of other related phenomena.

Presently, the ice budget of Antarctica is supposed to be slightly positive [+ 5 cm per year?]. The question of the ice budget could be settled by a polar orbiting satellite equipped with a Laser altimeter system of narrow footprint and with a range precision of ± 10 cm. Such a system could detect even gradual changes in ice surface elevation and slope, providing us with a direct measure of the volume change of the Antarctic ice sheet. In addition, this satellite could carry a multi-frequency microwave imager, yielding sea ice concentration, extent and type, sea surface temperature, and ice sheet ablation and accumulation. All these parameters are closely correlated to world climate.

These projects should be initiated and directed by an interdisciplinary study-group of IUGG.

REFERENCES

Bretterbauer K (1982) Eismassenänderungen und eustatisches Meeresniveau. Festschrift Pillewizer, Geowiss. Mitt. 21, Technical University Vienna, pp. 17-31

Farrell WE, Clark JA (1976) On Postglacial Sea Level. Geoph. J. R. Astr. Soc. 46, London, pp. 647-667

Kivioja LA (1967) Effects of Mass Transfers Between Land-Supported Ice Caps and Oceans on the Shape of the Earth and on the Observed Mean Sea Level. Bull. Geod. 85, Paris, pp. 281-288

Marcinek J (1977) Die Erde im Eiszeitalter. Geographische Bausteine, 19, VEB Hermann Haack, Leipzig

Nakiboglu SM, Lambeck K (1981) Deglaciation Related Features of the Earth Gravity Field. Tectonophysics, 72, Elsevier, Amsterdam, pp. 289-303

Schwarzbach M (1974) Das Klima der Vorzeit. Ferdinand Encke, Stuttgart

Sigl R (1975) Geodätische Astronomie. Sammlung Wichmann, 7, Karlsruhe, p. 65

Stolz A (1976) Changes in the Position of the Geocenter Due to Variations in Sea Level. Bull. Geod. 50, Paris, pp. 159-168

Glacial Rebound and Sea-Level Change: An example of Deformation of the Earth by Surface Loading

K. Lambeck
Australian National University
Australia

ABSTRACT:

One of the more important illustrations of the earth's response to external forcing is provided by the glacial rebound problem. The melting of the Late Pleistocene ice sheets and the distribution of meltwater into the oceans provides a time dependent and spatially variable surface load. A principal observation of the response of the earth to this change is in the relative shift in position of sea-level and the crust. Observations of this response provide constraints on the mantle rheology as well as on the glacial unloading process. This paper sketches out the problem, develops the sea-level equation and its solution and discusses the convergence requirements of the latter. It then discusses the principal characteristics of the global sea-level change and illustrates how earth and ice parameters can be separated by examining sea-level data in different areas, at different time intervals, and by using differential techniques. The final section summarizes some of the principal results and unresolved questions.

1. INTRODUCTION

Geomorphological observations have revealed a complex pattern of sea-level change for the past 18 000 years that are associated with the last deglaciation of the Late Pleistocene ice sheets. Near the centres of former glaciation the sea-level has dropped relative to the crust by hundreds of metres, while far from these centres, sea-level has risen by typically 130-150m. These observations are important for studying the tectonic histories of continental margins and ocean islands, for evaluating the mechanical response of the Earth to surface loading on time scales of 10^3-10^4 years, for constraining the melting histories of the large polar ice sheets, and for understanding present sea-level change.

The recognition that raised and warped shorelines in northern Europe reflected the Earth's response to glacial unloading go back to the nineteenth century and early attempts to quantify this response were made by Daly, Haskell, Niskanen, Vening Meinesz and others. The plate tectonics hypothesis, with the need to establish estimates of the mantle viscosity, resulted in a renewed interest in the problem and major contributions were made by McConnell (1968), Walcott (1972), Cathles (1975), Farrell and Clark (1976), and Peltier and Andrews (1976). Yet today, the problem is still not solved and continues to attract attention, in part because of the renewed interest in present and future sea-level change as a possible result of the enhanced Greenhouse effect. One reason for the lack of solution in the many faceted nature of the problem. Other than developing modelling procedures, models for the waxing and waning of the ice sheets need to be improved. The base of observational evidence for relative sea-level change also needs to be carefully evaluated and expanded upon.

The emphasis of this paper is on the geophysical modelling of the sea-level change produced by the melting of the ice sheets. Recent work has shown that in order to obtain realistic predictions a very high spatial resolution of the sea-level is required, whether this be at positions near the ice sheet or far from any centre of glaciation. Once achieved, these solutions begin to explain the complex patterns of change observed in a variety of regions; along the Australian shoreline, between nearby Pacific islands, or in northwestern Europe.

2. THE GLACIAL REBOUND PROBLEM

Consider a rigid planet partially covered by ocean. The sea surface will be an equipotential whose shape is determined by the gravity field of the planet and of the ocean waters. Extraction of water from the ocean to form an ice sheet leads to an overall lowering of the sea-level but not by a uniform amount because the redistribution of mass on the rigid planet's surface changes the shape of the equipotential surfaces.

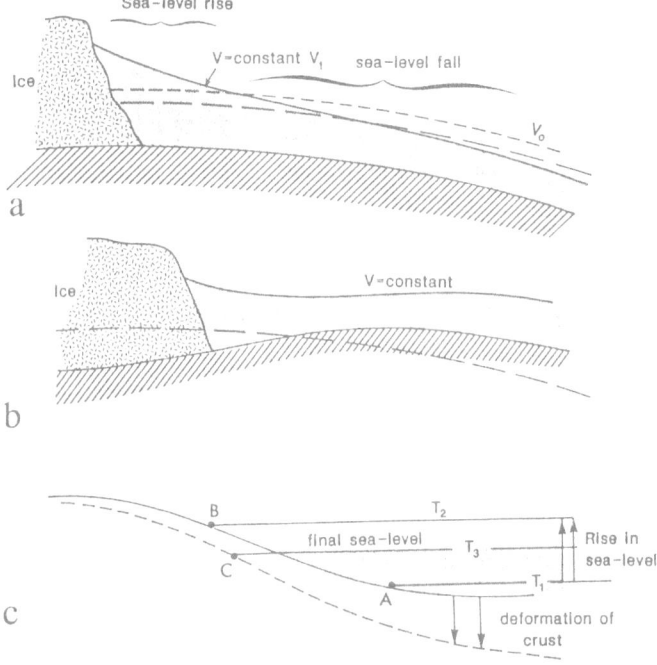

Figure 1:   Schematic illustration of the shape of the sea surface in response to
glaciation.
(a)   Near the edge of an ice sheet on a rigid planet. When the ice sheet
forms water is extracted from the ocean reducing the equivalent sea-
level, but because of the gravitational attraction of the ice and water
the sea-level change is not uniform.
(b)   Same as (a) but for a deformable Earth.
(c)   Far from the ice sheet along a continental margin. At time T_1 the sea
level begins to rise rapidly up to the time T_2 and the shore line moves
from point A to B. In this interval and afterwards the crust responds
to this loading by subsiding beneath the oceans and rising beneath the
continent, producing a tilting of the margin. Thus at time $T_3 > T_2$
sea-level actually appears to drop (point C).

Near the newly formed ice sheet the equipotential surface is pulled up and, depending
on the new mass distribution, sea-level may actually be higher than before. As melting
occurs sea-level drops in the vicinity of the ice sheet in response to the modification of
the equipotential surface and at the same time it rises as the volume of the ocean
increases (Figure 1a). On a deformable planet the growing ice sheet loads the crust and
induces flow in the mantle away from the loaded area while the reduced water load
produces a rebound of the oceanic crust.   The shape of the sea-surface is now
determined by the time dependent gravity field of the solid body as well as that of the
ice and water load (Figure 1b). Within and near the ice sheets, the response of the crust
is primarily determined by the change in the ice load. Further away, where typically

the fluctuations in sea-level are about 100 m, the crustal adjustments to the changing water load are of the order of 30 m.

Consider sea-level at a continental margin, far from the ice sheets. As meltwater is added into the oceans the sea floor is loaded and the sub-ocean mantle is stressed and mantle flow occurs from beneath the continent. The amount of deformation will be a function of the magnitude and spatial distribution of the water load, of the flow properties of the mantle, and of the mechanical properties of the lithosphere, but generally there will be a tilting of the margin (Figure 1c). At ocean islands the additional meltwater loads the sea floor uniformly about the island and the sea floor is depressed taking the island with it. But if the island is large, sub-lithospheric flow occurs from beneath the ocean to beneath the island and the later will tend to rise, resulting in a differential response between the island and the sea-floor.

What emerges from this heuristic description is a sea-level that displays complex spatial and temporal patterns of change. A zero order approximation of this change is provided by the time-dependent equivalent sea-level (esl), defined as

$$\zeta_{esl} = \zeta_o = \text{(change in ocean volume)/(ocean surface area)} \tag{1}$$

Spatial departures from this can be large and this definition has limited usefulness beyond being an estimate of the change in ocean volume. A more appropriate definition of the sea-level is

$$\zeta = \zeta_r + \zeta_i + \zeta_w \tag{2}$$

where ζ_r is the sea-level change that would occur on a rigid Earth, and includes ζ_{esl}. The second term ζ_i is the additional change produced by the response of the crust to the changing ice volume and ζ_w is the further adjustment produced by the response of the crust to the changing water load.

The relative importance of the three terms varies with time and position. It is therefore useful to discuss sea-level during the Late Pleistocene-Early Holocene, from about 20,000 to 8,000 years ago when ice melting occurred, and during the post-melting period of 8,000 years to the present. It is also useful to discuss sea-level for different regions: the far-field, the region far from the ice sheet where the primary contributions are ζ_{esl} and ζ_w; the near-field, being the region near the margin of the ice sheet where the dominant contribution comes from ζ_r and ζ_i ; the intermediate field between these two regions; and the region within the ice sheet margin where the dominant form is ζ_i. Characteristic sea-level curves can be established for each area (Figure 2) but considerable variation occurs within each zone and no simple zoning occurs.

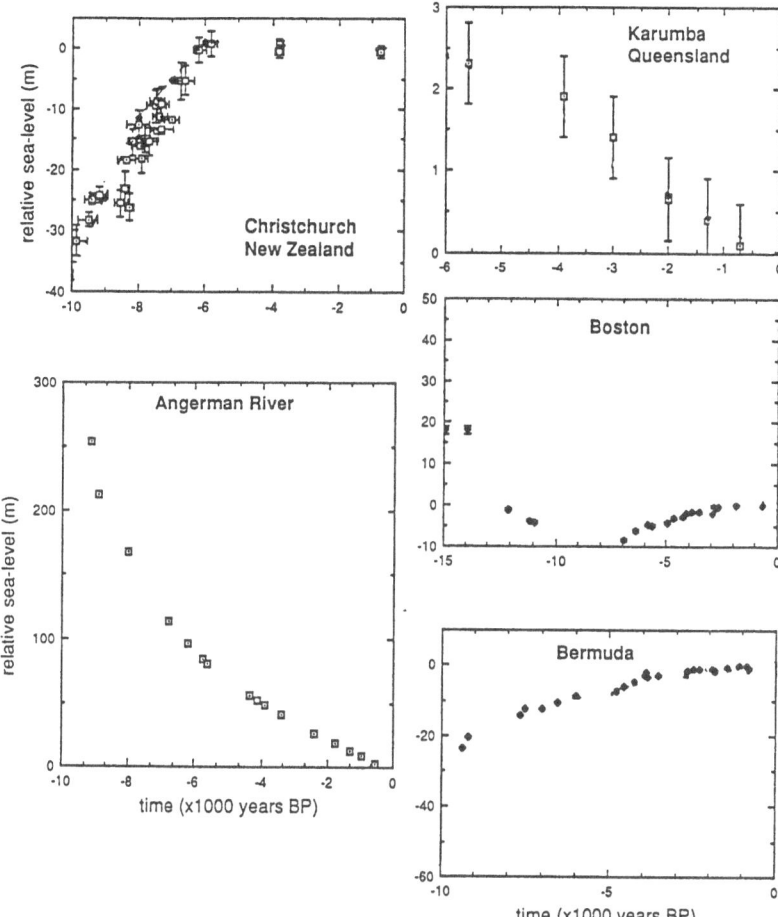

<u>Figure 2</u>: Characteristic sea-level observations

  (a) far-field in late Pleistocene and Holocene time (Christchurch, New
       Zealand)
  (b) far-field in late Holocene time (Karumba, Gulf of Carpenteria,
       Queensland)
  (c) near-field (Boston, Massachusetts)
  (d) intermediate-field (Bermuda)
  (e) site near centre of former Fennoscandian ice sheet (Angerman River,
       Gulf of Bothnia).

3. THE SEA LEVEL EQUATION

The sea-level variation can be defined by using a Green's function formulation for the change in gravitational potential due to the redistribution of the surface load. For a rigid Earth, the potential at a point **r** due to the ice load **r'** is given by

$$V_i(\mathbf{r}) = \int_{A_i} \mathcal{G}v\ (|\mathbf{r} - \mathbf{r}'|)\ \rho_i\ \zeta_i\ (\mathbf{r}')\ dA \tag{3}$$

where ζ_i is the ice height at $\mathbf{r}'$, ρ_i is the ice density, $\mathcal{G}v$ in the potential Green function

$$\mathcal{G}v = \frac{G}{R} \sum_{n=0}^{\infty} P_n(\cos\ \psi) \tag{4}$$

where P_n are Legendre polynomials and ψ is the angular separation of $\mathbf{r}$ and $\mathbf{r}'$. The integral is over the surface area of the ice. The radial shift in the equipotential surface due to the ice load is V_i/g or $(3\rho_i/4\overline{\rho})\ \Sigma_n\ I_n$ where

$$I_n = R^{-2} \int_{A_i} \zeta_i(\mathbf{r}')\ P_n\ (\cos\ \psi)\ dA \tag{5}$$

and $\overline{\rho}$ is the mean density of the Earth. The change in sea-level caused by this potential alone is

$$\zeta' = \{V_i - \langle V_i\rangle_0\}/g \tag{6}$$

where $\langle\ \rangle_0$ denotes the average value taken over the ocean. This is

$$\zeta' = -\frac{3}{4\pi}\ \frac{\rho_i}{\rho_0}\ \sum_n (I_n - \langle I_n\rangle_0)\ . \tag{7}$$

To this must be added the meltwater flowing into the ocean

$$\zeta'' = \{\zeta_0 + V_w - \langle V_w\rangle_0\}/g \tag{8}$$

where ζ_0 is the equivalent sea-level (c.f. eqn 1)

$$\zeta_0 = -M_i\ /\ 4\pi\ R^2\ 0_{100}\ \rho_w \tag{9}$$

where M_i is the change in total ice mass, considered to be positive for a growing ice sheet. $0_{100} \simeq 0.70$ is the zero degree term in the ocean function expansion. V_w is the potential of the water load, or analogous to (3),

$$V_w(\mathbf{r}) = \int_{A_0} \mathcal{G}v(|\mathbf{r} - \mathbf{r}'|)\ \rho_w\ \zeta(\mathbf{r}')\ dA \tag{10}$$

where ζ is the sea-level change and the integral is over the ocean area A_0. Then the final change in sea-level is, still for the rigid Earth model,

$$\zeta = \zeta_r = \zeta_0 - \frac{3}{4\pi} \frac{\rho_i}{\rho_0} \sum_n (I_n - \langle I_n \rangle_0) + \frac{3}{4\pi} \frac{\rho_w}{\rho} \sum_n (J_n - \langle J_n \rangle_0) \qquad (11)$$

with

$$J_n = R^{-2} \int_{A_0} \zeta(r') P_n (\cos \psi) \, dA \quad . \qquad (12)$$

This equation defines the change in sea-level with time and position on a rigid Earth due to the melting of ice and the addition of this meltwater into the ocean. The sea-level change ζ enters on both sides, on the right hand side through the integral J_n, and the equation is an integral equation that is solved by an iterative procedure. Mass is conserved and the sea-level remains an equipotential surface throughout.

For an elastic planet the Green's functions become

$$G_V = \frac{G}{R} \sum_{n=0}^{\infty} (1 + k_n' - h_n') P_n (\cos \psi) \qquad (13)$$

where k_n', h_n' are the load Love numbers. The sea-level change on the elastic planet now is (c.f. eqn 2)

$$\zeta_e = \zeta_r + \delta\zeta_e \qquad (14)$$

where the elastic contribution is

$$\delta\zeta_e = -\frac{3}{4\pi \, \overline{\rho}} \left[\rho_i \sum_n (I_n' - \langle I_n' \rangle_0) \, r - \rho_w \sum_n (J_n' - \langle J_n' \rangle_0) \right] \qquad (15)$$

with

$$I_n' = R^{-2} \int_{A_i} \zeta_i(r') (k_n' - h_n') P_n (\cos \psi) \, dA \qquad (16)$$

and a similar definition for J_n'. On a viscoelastic planet

$$\zeta_{ve} = \zeta_r + \delta\zeta_{ve} \qquad (17)$$

where, by using the correspondence principle (e.g. Peltier 1974),

$$\delta\zeta_{ve} (\varphi,\lambda;t) = L^{-1}\{\delta\zeta_e(\varphi,\lambda:s) \qquad (18)$$

and where L is the Laplace transform, L^{-1} is the inverse transform, and s is the Laplace transform parameter of dimension time^{-1}. Then $\delta\zeta_e(s)$ is of the same form as (15,16) but in which the I_n' and J_n' are replaced by their transforms, for example,

$$I_n' \Rightarrow \hat{I}_n' = R^{-2} \int_{A_i} \hat{\zeta}_i (\hat{k}_n' - \hat{h}_n') P_n (\cos \psi)\, dA \tag{19}$$

where $\hat{k}_n'$, $\hat{h}_n'$ are the load Love numbers in the Laplace transform domain. The resulting sea-level equation can now be written as (c.f. equation 2)

$$\zeta = \zeta_r + \zeta_i + \zeta_w \tag{20}$$

where

$$\zeta_i = Z_1 - \langle Z_1 \rangle \quad , \quad \zeta_w = Z_2 - \langle Z_2 \rangle \tag{21}$$

with

$$Z_1 - \frac{3}{4\pi} \frac{\rho_i}{\rho} \int_{A_i} L^{-1} \left[(\hat{k}_n - \hat{h}_n)\, \hat{\zeta}_i(s) \right] P_n (\cos \psi)\, dA_i) \tag{22}$$

and

$$Z_2 - \frac{3}{4\pi} \frac{\rho_w}{\rho} \sum_n \int_{A_o} \left[\int_{A_i} L^{-1} \left[(\hat{k}_n - \hat{h}_n)\, \hat{\zeta}_i(s) \right] dA_i \right] P_n (\cos \psi)\, dA_o \tag{23}$$

where the summation over n is from 0 to ∞, at least in principle. This sea-level equation is essentially that of Farrell and Clark (1976), and has been used by Peltier and Andrews (1976), Wu and Peltier (1983), Nakiboglu et al. (1983), and Nakada and Lambeck (1987).

Solutions of the sea-level equation require a knowledge of the evolution of the ice loads $\zeta_i(\varphi,\lambda:t)$ through space and time and of the viscosity structure of the planet through the Love numbers $(\hat{k}_n' - \hat{h}_n')$. Expressions for the latter can be derived for simple Earth models (e.g. Lambeck, 1988, p. 67) but for realistic solutions complete Earth models are essential. The elastic parameters and density are given by the seismic models but the viscosity is largely unknown and to be determined from the observations. Figure 3 illustrates these Love numbers in the Laplace transform domain for the elastic Earth model of Dziewonski and Anderson (1981), a constant mantle viscosity of 10^{21} Pas s, and a high viscosity (10^{25} Pa s) lithosphere of 50 km thickness. For large s, corresponding to small t, the response approaches that of an elastic planet.

For small s ($s \lesssim 10^{-4}$ a^{-1}) the response approaches that of an elastic shell over a fluid mantle in which all load stresses have relaxed. Further evolution of the response occurs with smaller s as stresses relax in the lithosphere. These results can therefore be interpreted qualitatively as a relaxation spectra for harmonic loads. Figure 4 illustrates the corresponding time dependent Love numbers L^{-1} ($\hat{k}'_n$ - $\hat{h}'_n$) for a step function surface load. At time $t = 0$, the response is that of an elastic body and with time the response decays to a limit corresponding, in this case, to the elastic response of a 50 km thick elastic shell overlying a fluid interior. By $t = 20,000$ years, much of the load-induced stresses have relaxed within the viscoelastic mantle.

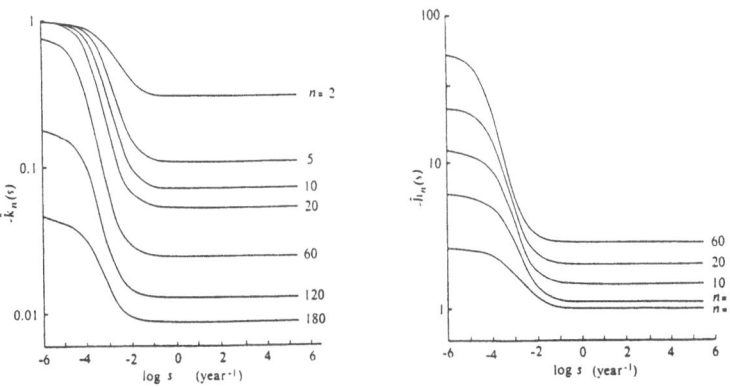

Figure 3: Load Love numbers $\hat{h}_n(s)$, $\hat{k}_n(s)$ as a function of the Laplace transform variable s (years^{-1}).

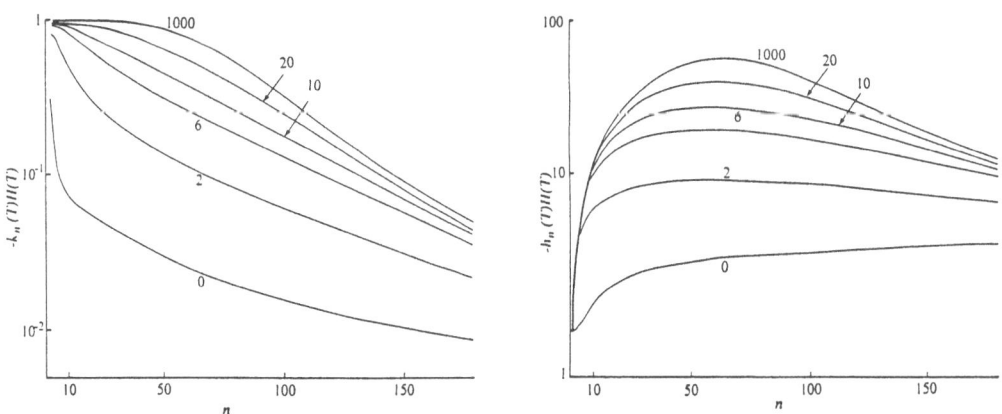

Figure 4: Load Love numbers k, h as a function of time (in units of 1000 years).

It has been shown by Peltier (1974) that the $(\hat{k}'_n - \hat{h}'_n)$ can be adequately expanded numerically into the form

$$(\hat{k}'_n - \hat{h}'_n) = \sum_{i=1}^{m} \frac{C_n^i}{S+S_i} + \left[k_e^n - h_e^n \right] \qquad (24)$$

where k_e^n, h_e^n are elastic load Love numbers. This is essentially an eigen function expansion in which the S_i correspond to the eigen values. This expansion is also used here in reaching solutions of equations (20)-(23).

The solution of the sea-level equation also requires a model of the temporal and spatial distribution of the ice load. The observed quantities are the limits of the ice sheet as the last glaciers receded and this is complemented with mechanical models for the ice sheets (e.g. Paterson, 1972; Hughes 1981). A critical unknown in this calculation is the amount of isostatic compensation that has occurred beneath the load. The effect of the delay in the isostatic response is another vexing question and it means that the ice models are not free from assumptions about the Earth's response to loading. Another major assumption in these reconstructions is whether the ice sheets are in static equilibrium or whether dynamic non-equilibrium models are more appropriate in which the ice volumes are driven by external climatic forcing and feedback mechanisms (Budd and Smith, 1981, 1987). The direct observations of the isochrones are usually restricted to the last 20,000 years and the earlier record is less clear, often inferred indirectly from palaeoclimatic or isotopic signatures or from the variations in sea-level itself (e.g. Chappell and Shackleton, 1986).

In the present formulation the ice sheets are described by ice columns of defined areal extent and whose height varies linearly with time in the interval t_j and t_{j+1} so that for column m and time t

$$\zeta_{mj}(t) = a_j t + b_j \qquad\qquad t_j \le t \le t_{j+1} \qquad (25)$$

The entire melting history is defined by a total of m such columns over r time intervals. The corresponding Laplace transforms are then convolved with the numerical transforms (24) of the Earth's response in order to evaluate the integrals Z_1, Z_2 (equations 22, 23) and the sea-level change (equation 20). Details are given in Nakada and Lambeck (1987).

4. CONVERGENCE REQUIREMENTS

If high resolution solutions of the sea-level equation are sought then the geometry of the surface loads must be known with very considerable detail. From eqns (20)-(23) the change in the sea-level with respect to the present day value $\zeta(t_o)$ can be written as

$$\Delta\zeta(t) = \zeta(t) - \zeta(t_0) = \sum_{n=0}^{N} \Delta\zeta^{(n)}(t) \qquad (26)$$

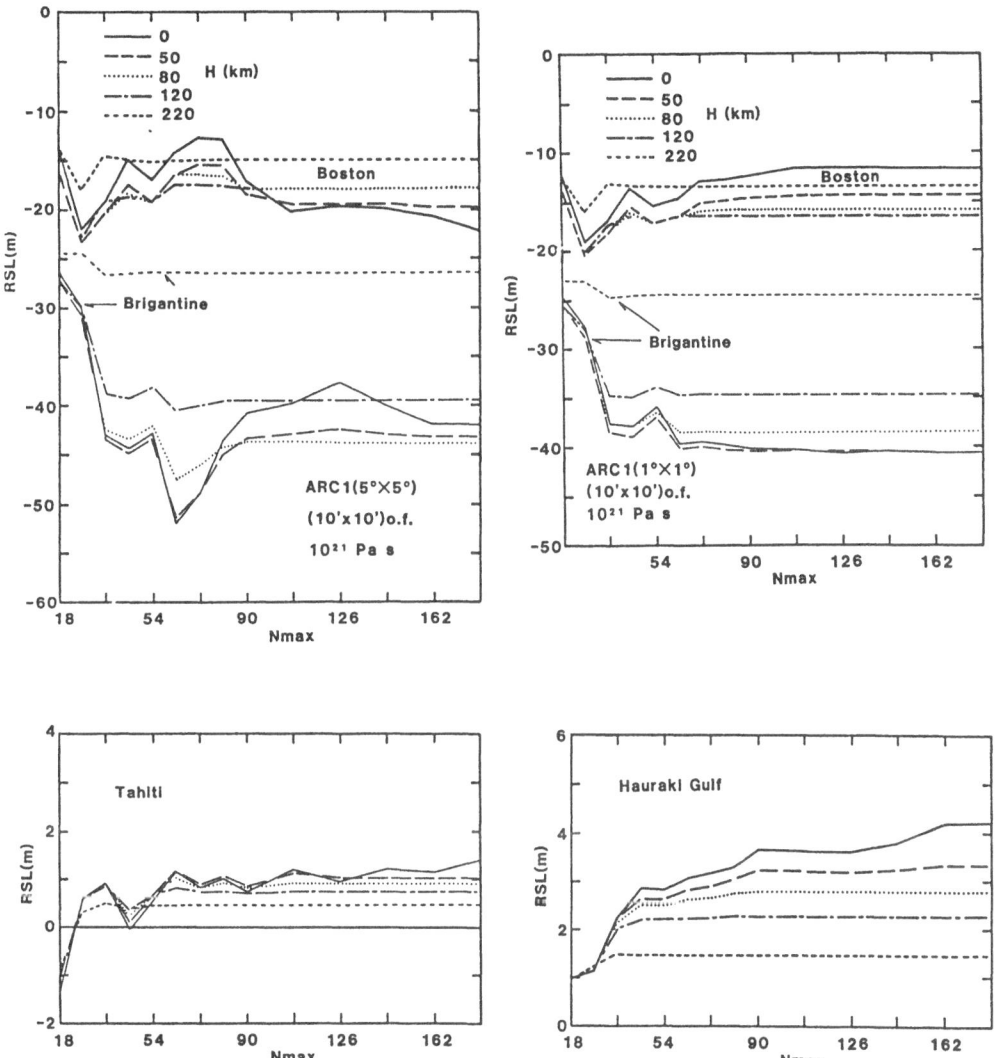

Figure 5: (a) Relative sea-level change $\Delta\zeta(t)$ as a function of N_{max} (equation 26) for a 5° resolution of the ice model at a site near the ice sheet margin.

(b) Same as (a) but for the 1° ice model.

(c) Same as (a) but for the far-field site of Tahiti.

(d) Same as (c) but for a site on the Hauraki Gulf of New Zealand.

Figure 5 illustrates $\Delta\zeta(t)$ for different degrees N_{max} of truncation of the expansion. In Figure 5 a comparison is also made at sites near the edge of the ice sheet for both the 5° and 1° descriptions of the the ice load. What these results indicate is that (i) convergence of the expansion can be very slow, particular for the coarse definition of the ice load, (ii) the value of $\Delta\zeta_i$ at convergence is not the same for the two loads because they define different mass distributions in the vicinity of the site, (iii) the convergence can be accelerated by introducing Earth models with thick lithospheres. Figures 5c and 5d illustrate the convergence at two sites far from the ice sheet and here the dominant factor is the description of the nearby water load. The island of Tahiti has a diameter of about 40 km and the meltwater load corresponds approximately to an infinite plate with a hole in it. If the resolution of the ocean function cannot resolve the island then the island moves up or down with the sea floor but if the island is resolved by the ocean function then some differential movement occurs as mantle material beneath the ocean is forced beneath the island. The maximum n required in the expansion is a function of lithospheric thickness and exceeds about 90 for $H = 50$ km. More extreme is the result for the Firth of Thames on the Hauraki Gulf of New Zealand, a shallow sea of about 100 by 150 km dimensions. In all examples the convergence is accelerated if the lithospheric thickness is increased but convergence is not on the same value in each instance.Because of the requirement of these very high, resolutions of the load, it may appear that a better approach than spherical harmonic expansions is to use a finite element formulation with a grid size that varies with distance from the point at which the sea-level response is to be evaluated. However when this response is evaluated globally for a large number of sites then the harmonic expansion solutions are the most convenient, provided that the convergence requirements are met. In the following solutions the ocean function is defined with a 10' resolution and expanded into spherical harmonics up to degree 180. The ice sheets are modelled with a 1° resolution and also expanded out to this maximum degree. For sites near the ice margin this is still not adequate but generally the detailed information required to evaluate the response at these sites with high accuracy is not available in any case.

5. CHARACTERISTICS OF THE THEORETICAL SEA-LEVEL CURVE

Equation (20) is solved for the three components $\zeta_r, \zeta_i, \zeta_w$ using ice models for the Arctic and Antarctic ice sheets. The Arctic model includes the Laurentide and Cordilleran ice domes and the Fennoscandian ice dome as defined by the ICE1 model of Peltier & Andrews (1976) but smoothed and defined with 1° resolution (the ARC1 model of Nakada and Lambeck 1987, 1989). The 1° resolution model ARC3 includes an ice sheet over the Barents and Kara Seas of the northwestern Soviet Union. The Antarctic model ANT3 is a maximum ice model discussed by Nakada & Lambeck (1987, 1989).

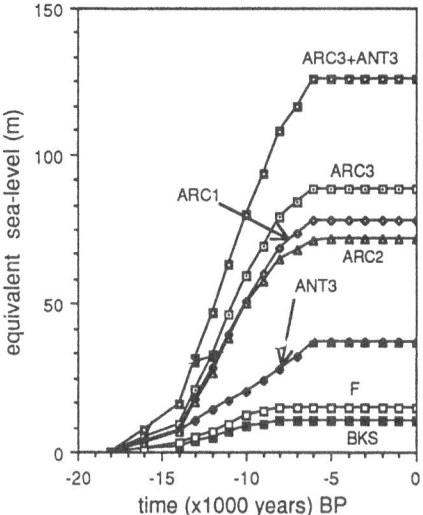

Equivalent sea-level curves for the Arctic ice models ARC1, ARC2, ARC3 and the Antarctic ice model ANT3. ARC1 is a 1° interpolated version of model ICE1 by Peltier and Andrews (1976). ARC2 corresponds to an interpolated version of ICE2 of Wu and Peltier (1983) which represent an adjustment of ICE1 to better fit the observational data. This model is therefore a function of the Earth model adopted by them. ARC3 is ARC1 plus a schematic model for the Barents-Kara ice sheet (BKS). All Arctic models include a Fennoscandian ice dome (F). AN3 is from Nakada and Lambeck, 1987, 1988).

The zero order approximation to the sea-level change resulting from the melting of these ice models is the time dependent equivalent sea-level curve (esl) defined by equations (1) and (9) and this function is illustrated in Figure 6. This curve can be approximately constrained by sea-level observations between 18 000 and 6000 years ago at far-field sites (Nakiboglu et al. 1983; Nakada and Lambeck 1988) and the adopted ice models are consistent with these observations. A number of different Earth models covering a range of plausible viscosity profiles are used and the appropriate parameters are discussed in Table 1. The mantle has been divided into two layers: an upper region, extending from the base of the lithosphere to the 650 km seismic discontinuity, and a lower mantle.

The solutions are expressed either relative to sea-level at the onset of glaciation, $\zeta(t)$ as defined by equation (23), or relative to the present level, $\Delta\zeta(t)$ as defined by equation (26). This latter quantity can also be expressed as the sum of 3 terms as

$$\Delta\zeta(t) = \Delta\zeta_r + \Delta\zeta_i + \Delta\zeta_w \qquad (27)$$

Table 1: Viscosity parameters for different Earth models used in forward modelling. All models use realistic elastic parameters and density profiles with the depth according to the model of Dziewonski and Anderson (1981).

Model	Viscosity (Pa s)	
	η_{um}	η_{lm}
E1	10^{20}	10^{21}
E2	2×10^{20}	10^{21}
E4	10^{21}	10^{21}
E14	2×10^{20}	10^{22}
E16	10^{21}	10^{22}
E28	10^{21}	10^{23}

where $\Delta\zeta_r$ defines the relative sea-level change on a rigid Earth, $\Delta\zeta_i$ in the additional contribution produced by the rebound of the crust as ice unloading proceeds, and $\Delta\zeta_w$ is the additional contribution produced by the deformation of the Earth in response to the water loading. Relative sea-levels have been predicted for a number of sites, including sites within the limits of the former ice sheets (Angerman River, Gulf of Bothnia, Northern Europe and Cape Henrietta Maria, Hudson Bay, Canada), in the near-field, near the margins of these ice sheets (e.g. Boston and the nearby sites of Barnstable, Massachusetts), in the intermediate field such as Bermuda or The Hague (Netherlands), and in the far-field in Australia and the South Pacific.

The rigid body term ζ_r. The rigid term ζ_r represents the effect of the rise or fall in sea-level resulting from deglaciation or glaciation on a rigid Earth with the requirement that the new sea surface remains an equipotential at all times. Within the ice sheet itself the shift of the equipotential surface is the sum of two opposing contributions: as melting proceeds the equipotential drops because of the reduced gravitational attraction of the ice load but at the same time it rises because the water level rises as melting proceeds. (Imagine the case of a vertical hole through the ice with a connection to the open sea. In reality the result has only sense from the time onwards when the site is exposed to the open sea.) At the Angerman River site in the Gulf of Bothnia, near the centre of the former Fennoscandian ice dome, the two parts contributing to ζ_r are almost equal but of opposite sign, with the result that ζ_r is nearly zero. For sites within the larger Laurentide ice sheet, the gravitational term exceeds the rise in sea-level term and the equipotential falls as melting proceeds (see results for Cape Henrietta Maria, Figure 7). Near the edge of the ice sheet the rise in water level exceeds the gravitational attraction of the ice and the net result is a rise in sea-level but at a rate that is less than the rate at which melt-water is added into the oceans. Further away from the ice sheet the rigid term is primarily one of a rise in sea-level from 18000 to about 6000 years ago as the melt water is added into the ocean. until it approximates the equivalent sea-level curve. Near the Hawaiian ridge, $\zeta_r \lesssim \zeta_{esl}$ but further south $\zeta_r \gtrsim \zeta_{esl}$.

The ice unloading term ζ_i. The term ζ_i is produced by the Earth's non-rigid response to the change in the ice load and it will be a function of the evolution of the ice load geometry with time as well as of the Earth's response function. At sites within the ice sheet the ζ_i term is one of an apparent lowering of sea-level as the crust emerges after unloading. At Cape Henrietta Maria, for example, ζ_i has a predicted amplitude of about 700 m for the ice-earth model ARC3-E4 (Figure 7a). The ice thickness at maximum glaciation was about 3000 m and the maximum rebound would be about 1200 m if a state of local isostasy had been achieved by the time the ice sheets reached their maximum extent. Observational estimates are not available of the maximum rebound that has occurred since the commencement of deglaciation, but observations at later times indicate that the rebound is less than predicted from the simple isostatic models. This is a result of several factors; (i) of the lithosphere possessing a finite strength so that part of the load is supported by the flexural stresses within this layer and the isostatic compensation is regional rather than local, (ii) of regional isostasy not having been achieved at the time of onset of deglaciation, because of the mantle viscosity, and (iii) of the rebound not yet being complete.

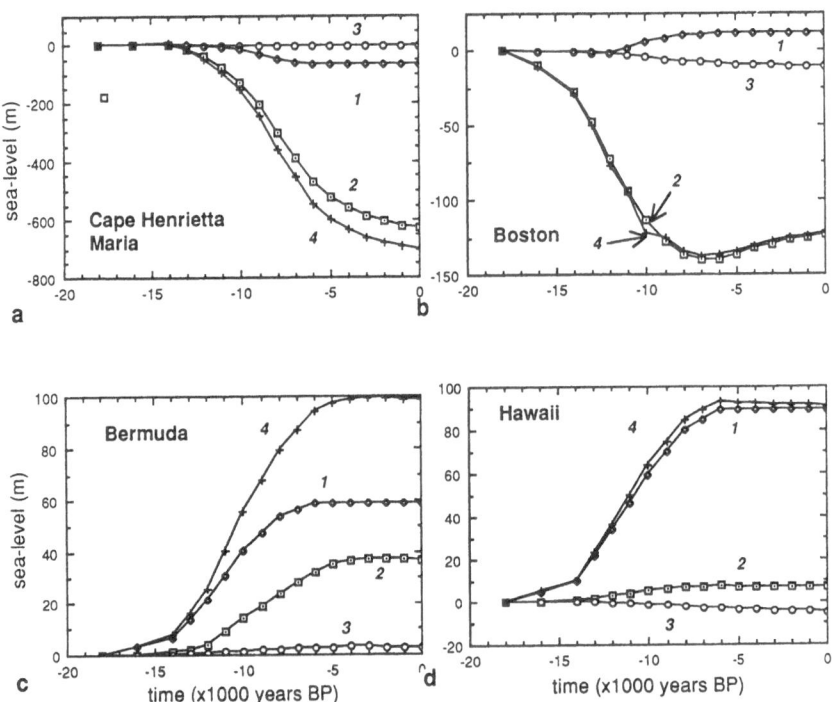

Figure 7: Predicted contributions to the sea-level change as defined by equation (20) at four different sites. Curve 1 corresponds to ζ_r, curve 2 to ζ_i, curve 3 to ζ_w and curve 4 to the total ζ. All change is with respect to sea-level at the onset of deglaciation at 1800 years ago.

Near the margins of the ice sheet the terms ζ_i and ζ_r are generally of opposite sign, ζ_r rises as melting proceeds whereas ζ_i falls, but ζ_i generally exceeds ζ_r in magnitude and the sea-level change from the combined effect is primarily one of an apparent falling level (Figure 7b). The ice unloading term here is spatially variable over quite short distances in the vicinity of the former ice margins and the predictions are about equally sensitive to mantle rheology and ice model parameters. Further from the ice front, at sites free from past localized ice loading, the two terms are of comparable magnitude

(Figure 7c) although at some sites the sign of ζ_i changes with time: initially the predicted ζ_i is an apparent fall in sea-level as the crust rebound to the early stage of unloading but later, as the ice retreats further from the site, an apparent rise in sea-level is predicted and this amplifies the ζ_r term.

The water loading term ζ_w. Compared to ζ_r and ζ_i the melt-water term ζ_w is small in the near-field (Figure 7), rarely exceeding 20 m for the combined ARC3 and ANT3 ice models. In the far-field after termination of deglaciation, ζ_w is generally larger than ζ_i (Figure 8). In a first approximation the time dependence of the ζ_w term is proportional to ($\zeta_r + \zeta_i$) and at sites far from the ice sheet this dependence follows closely the equivalent sea-level curve. The amplitude of ζ_w is a function of the load distribution in the vicinity of the site and is therefore a function of the coastal geometry as well as of ($\zeta_r + \zeta_i$) and the term may vary spatially over even relatively short distance if the coastline geometry is complex (Figure 8). In contrast ζ_i remains relatively constant over a given area of the far-field. Differential values of the sea-level between nearby sites n,m far from the ice sheets,

$$\delta\zeta_{nm} = \Delta\zeta_n - \Delta\zeta_m, \tag{28a}$$

are largely independent of the other two contributions and reflect primarily the response of the mantle beneath the sites to the change in water load through time, or

$$\delta\zeta_{nm} \cong \Delta\zeta_{w(n)} - \Delta\zeta_{w(m)}. \tag{28b}$$

Relative sea-level curves. The relative sea-level, as defined by (26) is the sum of the three terms $\Delta\zeta_r, \Delta\zeta_i, \Delta\zeta_w$ each of which varies spatially and temporally in a complex manner. The relative sea-levels are likewise predicted to exhibit considerable variability and no single curve adequately characterizes the sea-level change. Predicted sea-levels are a function of both the ice and earth-response parameters but, as illustrated by the above examples, this dependence varies with geographical location and time. This means that by comparing model predictions with observations of sea-level change it does become possible to separate these parameters. Figures 9-14 illustrate a range of predictions for sea-level change as a function of mantle viscosity, lithospheric thickness, and ice models which demonstrate that the separation of parameters can be achieved if adequate observational data can be found. For the period up to about 6000 years ago, the relative sea-level change in the far-field is insensitive to mantle parameters within a broad range of viscosities, lithospheric thickness, and details in ice loads. It is, however, strongly dependent on the equivalent sea-level rise; on the rates and timing of the melt-water addition (Figure 9). Particularly important is the sea-level at the time of onset of deglaciation as this constrains the total volume of melt-water. Far-field observations of sea-level, therefore, constrain the gross aspects of the Arctic and Antarctic ice models.

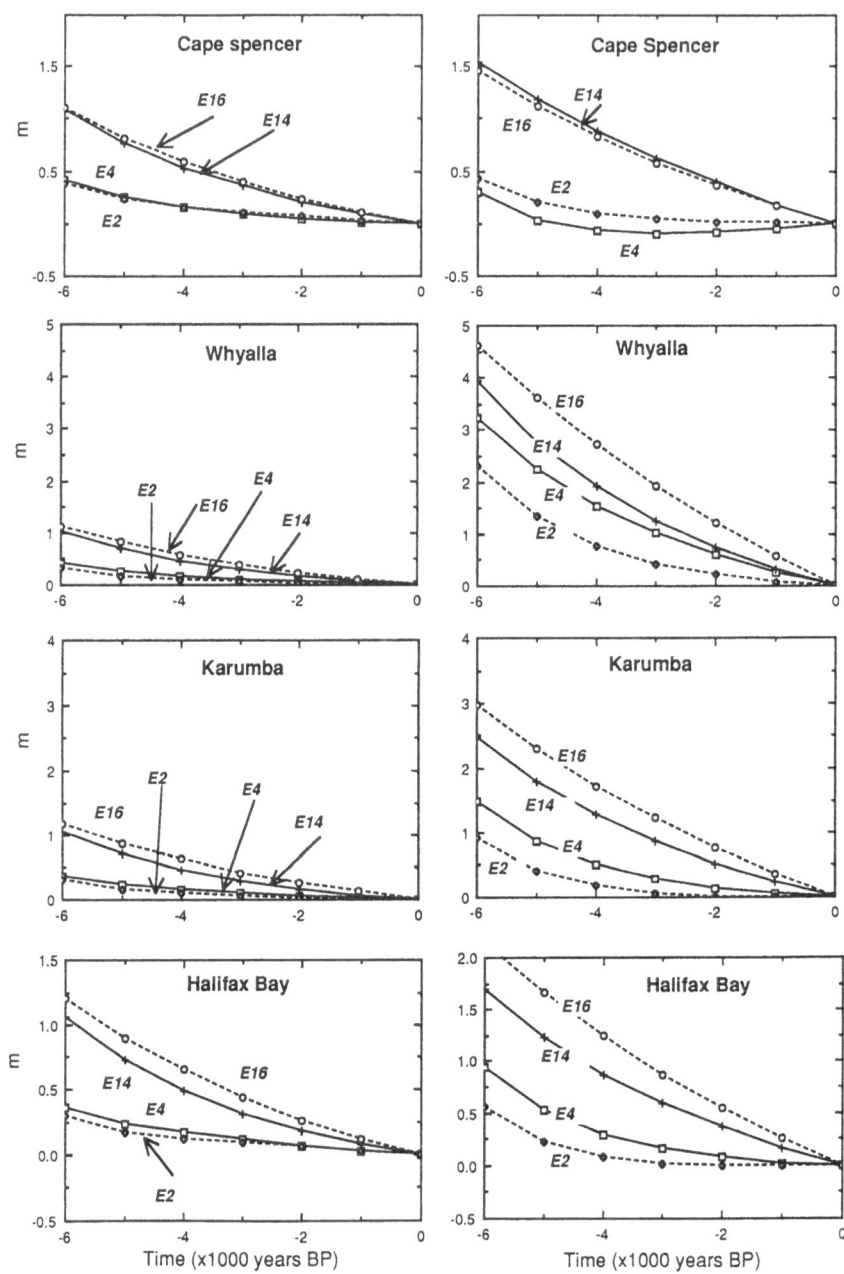

Figure 8: Contributions to relative sea-level at far-field sites during the Late Holocene. The $\Delta\zeta_i$ terms are given on the left and the $\Delta\zeta_w$ terms are given on the right for four Earth models (Table 1). (Note different scales used.) Whyalla and Cape Spencer are two sites about 300 km apart in the Spencer Gulf of South Australia and Karumba and Halifax Bay lie about 600 km apart on opposite sides of the Cape York Peninsula of Northern Queensland. Differential values of sea-level between nearby sites n,m far from the ice sheet

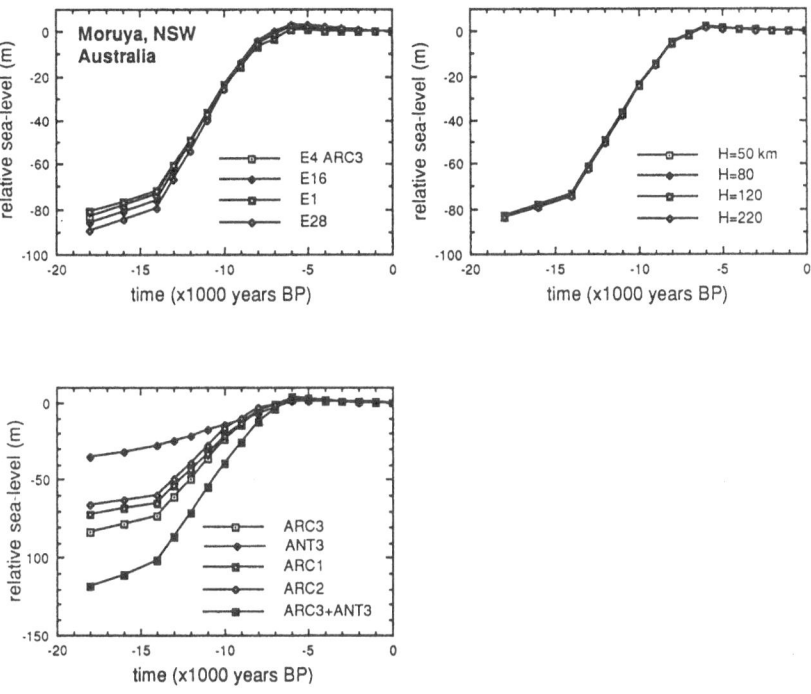

<u>Figure 9</u>:   Predicted sea-level change for past 18000 years at the far-field site of Moruya (New South Wales, Australia) as a function of mantle viscosity, lithospheric thickness and ice model.

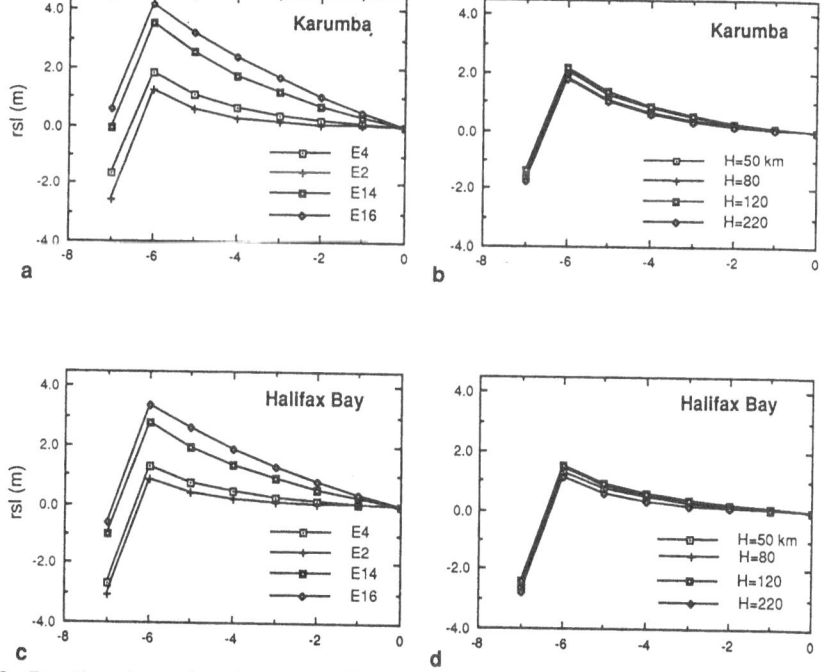

<u>Figure 10</u>: Predicted sea-levels at two far-field continental margin sites for past 7000 years.

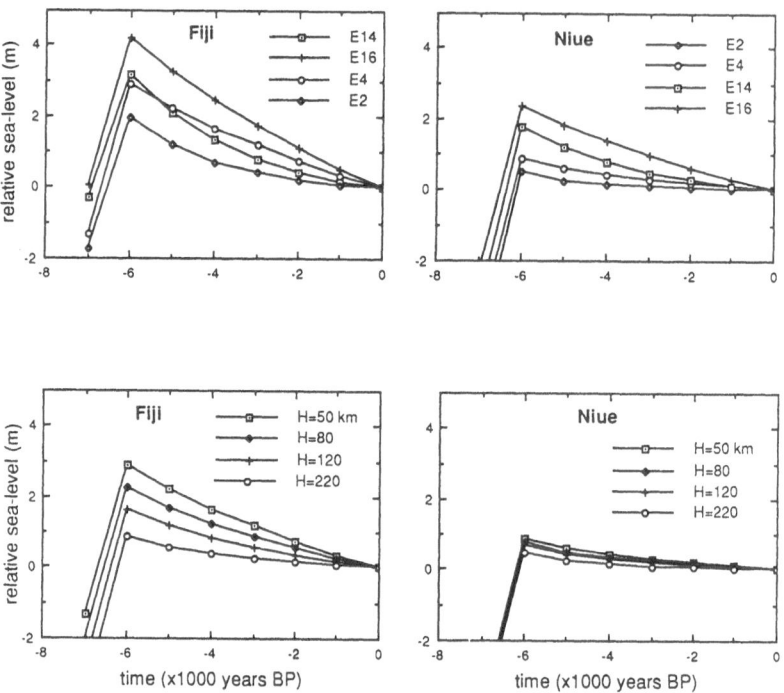

Figure 11: Same as Figure 10 but for island sites. The Fiji island of Viti Levu has a diameter of about 130 km compared with about 20 km for Niue.

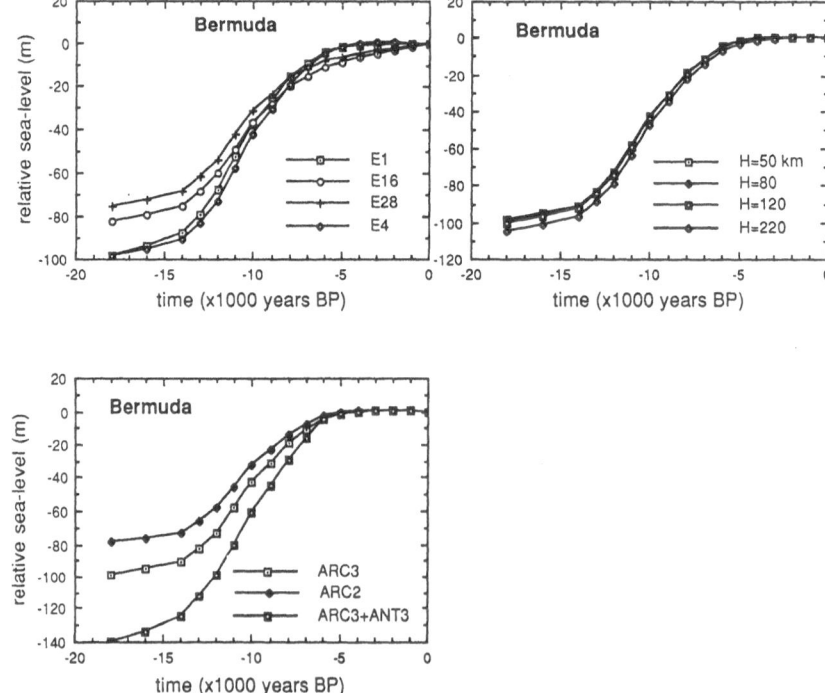

Figure 12: Same as Figure 9 but for a site in the intermediate field.

131

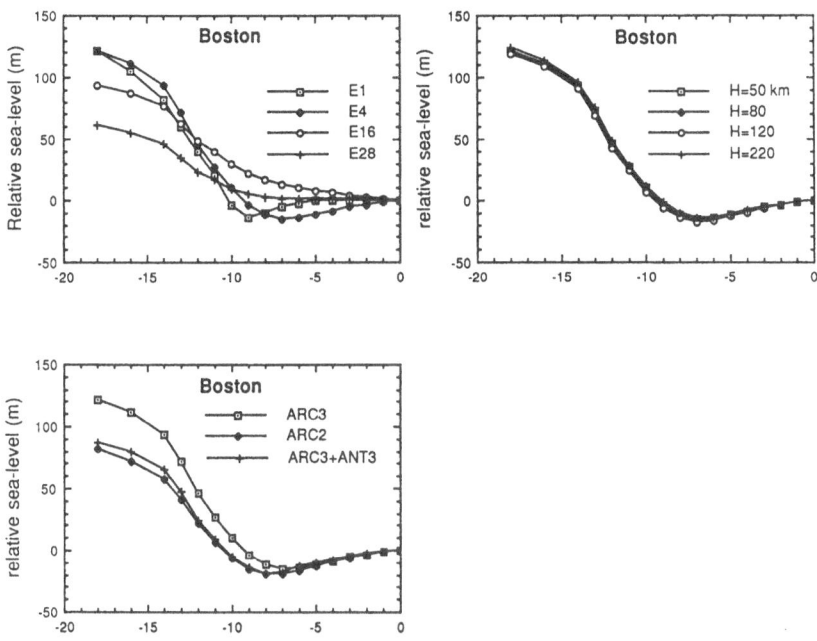

Figure 13: Same as Figure 9 but for a site near the ice margin.

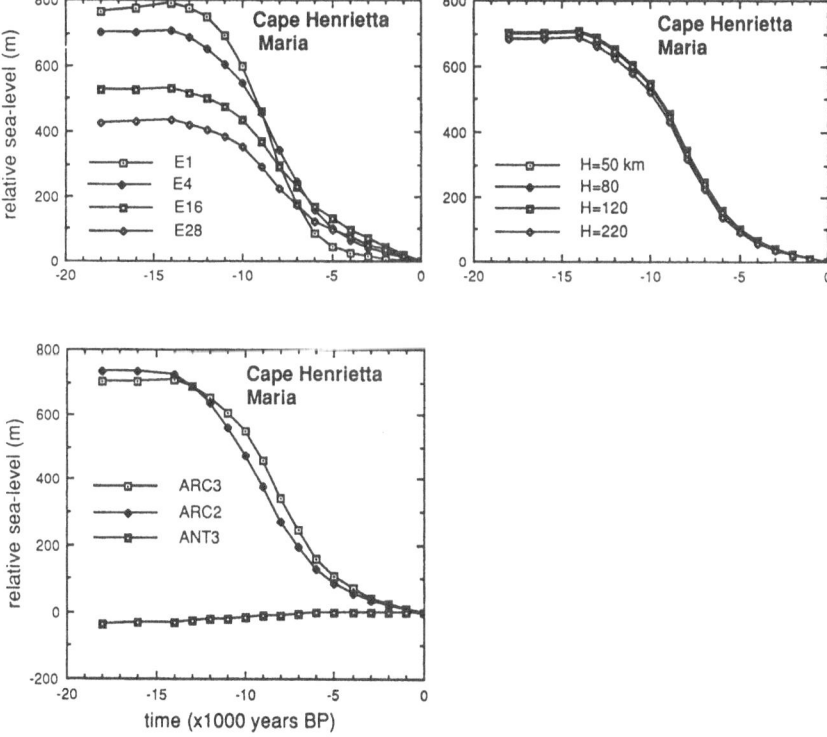

Figure 14: Same as Figure 13 but for a site near the centre of the Laurentide ice sheet (Cape Henrietta Maria).

The predicted far-field sea-levels along continental margins for the past 7000 years are characterized by a maximum value of up to a few metres above the present value at about 6000 years ago. A careful examination illustrates considerable regional variations in these amplitudes because of the melt-water term $\Delta\zeta_w$ which is strongly dependent on coastal geometry. This is illustrated in Figure 10. Here, spatial differential values, as defined by equation (28) provide a sensitive measure of Earth structure, largely free from assumptions made about the ice models (Nakada and Lambeck 1989). The absolute values of these highstands provide a means of fine-tuning the ice melting models in the past 6,000 years, once the Earth structure is established from the differential values. An important point is that these far-field spatial differential values are primarily sensitive to mantle structure near the sites, and not to global structure. This opens, therefore, the possibility of examining lateral variations in mantle viscosity. Important here are observations from ocean islands because predictions for nearby islands of different sizes indicate that differences in sea-levels are good indicators of mantle structure (Figure 11).

At sites in the intermediate field, such as Bermuda (Figure 12) the sea-level change is a function of both the rigid and ice terms and for past 6000 years the ice term $\Delta\zeta_i$ remains important. For models of relatively low viscosity (e.g. E1 and E4) the sea-levels are predicted to have reached their present value already several thousand years ago whereas for higher viscosities (e.g. models E16 or E28) sea-level will still be encroaching on the island (i.e. the island is subsiding).

Close to the ice margins the predicted sea-levels vary rapidly with positions relative to the ice front at the time of glaciation and major differences occur between nearby sites. The observations are particularly dependent on the assumed ice model (Figure 13) but also on mantle viscosity. Spatial differential values (equation 28) of sea-level may be useful here in separating out the two factors. Predictions for nearby sites, for example, are approximately independent of mantle viscosity unless the upper mantle value is $\gtrsim 5\times10^{20}$ Pa s and the lower mantle value is 10^{21} Pa s. Likewise, the differential values are approximately independent of lithospheric thickness unless this value is $\gtrsim 200$ km. However, because the available number of observations are usually small and the number of parameters required to model the ice sheet is large, it will generally not be possible to invert the differential observations to obtain unique ice models (Quinlan, 1981). For example, the ARC2 ice model has the same effect on sea-level at Boston as does the addition of the Antarctic ice sheet.

Within the limits of the former ice sheet margin $\zeta_i >> \zeta_r$, ζ_w and the relative sea-level curve is determined almost wholly by the response of the crust to the unloading of the surrounding ice sheet. Here the predicted rebound is relatively insensitive to the details of the ice models at the margins but it is sensitive to the thickness of the ice sheet (Figure 14). The horizontal dimensions of the Fennoscandian ice sheet are of the order of 1000-1500 km and the rebound is primarily sensitive to upper mantle viscosity. The rebound due to the much larger Laurentide ice sheet is, however, sensitive to the viscosity structure of the entire mantle.

6. PRINCIPAL CONCLUSIONS AND UNRESOLVED QUESTIONS

The predictions discussed in the preceding section and illustrated in Figures 9-14 do indicate that the sea-level change associated with the late Pleistocene melting can be expected to exhibit a complex global pattern. Observations from a variety of regions are generally consistent with this variability. In order to predict this change globally and with accuracy, it is necessary to have a detailed knowledge of the melting histories of the ice sheets, both spatially and temporally. Adequate information is generally not available from glaciological information alone. There remains, for example, debate about the volume of ice locked up in the ice sheets and about the extent of these ice sheets in the northern hemisphere. Estimates of the Late Pleistocene ice sheet over the continental shelf of the Barents and Kara Seas (north-western Soviet Union) range from nearly zero to a volume equal to the Fennoscandia ice sheet. Eastern Siberia has generally been assumed to have been free from major ice domes during the last glacial but recently it has been suggested that a Pleistocene ice sheet existed here that rivalled the Antarctic ice sheet. There remains a similar uncertainty about the extent of Late Pleistocene melting of the Antarctic ice sheet.

The examination of past sea levels can contribute significantly to establishing constraints on these ice volumes and on locating the centres of loading. For example, observations of sea-level in the far-field generally indicate that this level was 130-150 m below the present level (e.g. Chappell, 1987). The predictions of these levels at the far-field sites are not strongly dependent on mantle rheology and they approximate closely the equivalent sea-level estimate. This is not so for sites closer to the ice sheet such as Bermuda and Florida and observations from sites in the intermediate field may not be a reliable indicator of equivalent sea-level, and their use could lead to a systematic underestimation of past ice volumes.

The examination of sea-level change near the postulated ice domes can also provide an important way of testing these glacial hypotheses. Raised beaches in the vicinity of the Antarctic base Davis, for example, establish that rebound has occurred in Holocene time and this points to unloading of the Antarctic ice sheet. From a single site the volume of unloading cannot be established and further examination of the Antarctic shoreline and far southern islands is necessary. Similarly, a re-examination of the Eurasian Arctic shorelines may be rewarding.

The far-field observations of sea-level from 12000-6000 years ago establish the equivalent sea-level curve for the latter part of the last deglaciation cycle. A possible problem here, however, concerns the time scale. Ice models are constrained largely by radiocarbon dates but varve chronologies have also been used and the two may differ significantly. The sea-level data is constrained mainly by radiocarbon ages, but sometimes uranium-thorium dates are used. If radio-carbon ages are used for both ice and sea-level data the time scale is consistent but not necessarily uniform. Even the consistency can be challenged because generally the major corrections for fractionation and reservoir effects are not made and these can be significant if the dated materials represent different environments (marine or estuarine water, submarine or subaereal). Lambeck and Nakada (1989) discuss this problem further in the context of the northwestern European sea-level and rebound evidence. Within the limitations of this time scale question, the far-field sea-level observations indicate that substantial melting of Antarctic ice occurred in phase with, or possibly as much as 1000 years after, the Laurentide melting (Nakada and Lambeck, 1988b). If this can be substantiated, this presents evidence on how the Antarctic ice sheet responds to rising sea-level, an important question in evaluating possible future sea-level change.

The predictions for the past sea-levels in the far-field exhibit considerable variability in amplitude and timing of the occurrence of the highstand at about 6000 years ago. The amplitudes are small, ranging from a few tens of centimetres to a few metres, but the precision of the observations is also high (e.g. Chappell et al. 1982). The regional variation in these amplitudes provide important constraints in mantle structure (Nakada and Lambeck, 1989). Preliminary analyses suggest that the upper mantle has a viscosity that may be 50 times less than the lower mantle, and that regional variations in upper mantle viscosity occur, with continental mantle having a higher viscosity than oceanic mantle.

An important aspect of these predictions is that they indicate that very high resolution models are essential. Coastline geometry must be modelled with very high resolution and, in many localities, it is necessary to model the time dependence of the coastline as sea-level rose during Holocene time if the highstands are to be modelled with precision. High accuracy models of the ice sheet melting are also essential if sea-levels are modelled near the ice sheet margins. This information is generally not available and sea-level observations from near the ice margin are of limited value for modelling the Earth's response.

Another unresolved question concerns the assumption of the linear viscoelastic response for the Earth. Perhaps the only justification for this choice is that it is mathematically convenient. It means that the viscosity parameters should be considered as effective parameters only and their interpretation in terms of realistic physical parameters requires caution. However, no observational evidence at this stage warrants the introduction of a non-linear rheology for modelling glacial rebound, particularly at site far from the centres of ice unloading. Other modelling questions concern the question of whether the mantle is treated as adiabatic or non-adiabatic (Cathles, 1975; Fjeldskaar and Cathles, 1984). A further question concerns the effect of earlier cycles of glaciation and deglaciation (Wu and Peltier, 1983; Nakada and Lambeck, 1987). Clearly much work remains to be done before we can conclude that the glacial rebound problem is solved but the progress made over the past decade or so, and the importance of the goals that can be achieved, make further study worthwhile.

REFERENCES

Budd WF, Smith IN (1987) Conditions for growth and retreat of the Laurentide ice sheet. Geography. Phys. Quat. 41: 312-321.

Budd WF, Smith IN (1981) The growth and retreat of ice sheets in response to orbital radiation changes. In Sea Level, Ice and Climatic Change. Int. Asst. Hydrol. Sci. Publ 131: 369-409.

Cathles LM (1975) The Viscosity of the Earth's Mantle. Princeton University Press, New Jersey.

Chappell J (1987) Late Quaternary sea-level changes in the Australian region, in Sea-level Changes: 296-331, eds. Tooley MJ, Shennan I, Basil Blackwell, New York.

Chappell J, Shackleton, NJ (1986) Oxygen isotopes and sea-level. Nature 324: 137-140.

Chappell J, Rhodes EG, Thom BG, Wallensky EP (1982) Hydro-isotasy and the sea-level isobase of 5500 B.P. in north Queensland, Australia. Mar. Geol. 49: 81-90.

Dziewonski AM, Anderson DL (1981) Preliminary reference Earth model. Phys. Earth planet Int. 25: 297-356.

Farrell WE, Clark JA (1976) On postglacial sea-level. Geophys. J.R. astr. Soc. 46: 647-667.

Fjeldskaar W, Cathles LM (1984) Measurement requirements for glacial uplift detection of nondiabatic density gradients in the mantle. J. Geophys. Res. 89: 10115-10124.

Hughes TJ (1981) Numerical reconstruction of paleo-ice sheets in Denton & Hughes (1981): 222-261.

Lambeck K (1988) Geophysical Geodesy: The Slow Deformation of the Earth. Oxford University Press, 718 pp.

Lambeck (1989) Glacial rebound, sea-level change, and mantle viscosity (1989) Harold Jeffreys Lecture. Q.J. Roy. Astron. Soc. (in press).

Lambeck K, Nakaka M (1989) Glacial rebound and sea-level change in Northwestern Europe. Geophys. J. (subm.).

McConnell, RK (1968) Viscosity of the mantle from relaxation time spectra of isostatic adjustment. J. Geophys. Res. 73: 7089-105.

Nakada M, Lambeck K (1987) Glacial rebound and relative sea-level variations: a new appraisal. Geophys. J. R. astr. Soc. 90: 171-224.

Nakada M, Lambeck K (1988) The melting history of the late Pleistocene Antarctic ice sheet. Nature 333: 36-40.

Nakada M, Lambeck K (1989) Late Pleistocene and Holocene sea-level change in the Australian region and mantle rheology. Geophys. J. 96: 497-517.

Nakiboglu SM, Lambeck K, Aharon P (1983) Postglacial sea-levels in the Pacific: implications with respect to deglaciation regime and local tectonics. Tectonophys. 91: 335-358.

Paterson WSB (1972) The Physics of Glaciers. Pergamon, New York.

Peltier WR (1974) The impulse response of a Maxwell Earth. Res. Geophys. Space Phys. 12: 649-69.

Peltier WR, Andrews JT (1976) Glacial isostatic adjustment-I: The forward problem. Geophys. J.R. astr. Soc. 46: 605-46.

Quinlan GM (1981) Numerical models of postglacial relative sea-level change in Atlantic Canada and the Eastern Canadian Arctic. Thest, Dalhousie University Halifax.

Walcott RI (1972) Late Quaternary vertical movements in eastern North America. Rev. Geophys. 10: 849-884.

Wu P, Peltier WR (1983) Glacial isostatic adjustment and the free air gravity anomaly as a constraint on deep mantle viscosity. Geophys. J.R. Ast. Soc. 74: 377-449.

Geodetic Analysis of Motion at a Convergent Plate Boundary

W.I. Reilly
Geophysics Division
Department of Scientific and Industrial Research
Wellington, New Zealand

ABSTRACT:

A simplistic model of the superficial interaction between two lithospheric plates would comprise two rigid bodies separated by a linear discontinuity. Where continental crust straddles an obliquely convergent boundary, however, as in New Zealand, extensive deformation can be seen over a zone several hundred kilometres in width. Past analyses of repeated geodetic surveys have indicated that shear-strain may be occurring at a steady rate, seemingly in the absence of observed discrete movement on faults. Such strain is by no means homogeneous over distances of hundreds of kilometres, and its elucidation is aided by the use of continuum models of deformation, which can be introduced directly into the simultaneous adjustment of large repeated survey networks.

Within the Hikurangi convergent margin of New Zealand, shear strain tends to increase eastward (toward the subducting Pacific plate and the offshore trench), and to have its shortening direction (WNW/ESE) consistent with the occurrence of frictional slip at the plate interface. Notable departures from this pattern indicate the back-arc spreading of the North Island volcanic region, and, on the north-east coast, possible trenchward sliding of the rapidly uplifted sediments of the accretionary wedge. At the southern end of the subducting zone of the Hikurangi Margin, the transition to a transform-fault regime in the South Island is characterised by marked bending, concave to the north-west.

1 INTRODUCTION

The superficial kinematics of plate-boundary regions are most readily observed where the boundary passes through (or near) continental crust lying above sea-level. Patterns range from a simple tensional regime at a divergent margin (as in Iceland), through the strike-slip motion of a transform-fault region (as in California), to the more complex behaviour of convergent margins (e.g. the Mediterranean zone of southern Europe; Japan; New Zealand). Intraplate velocities in the range of 10 to 100 mm/year make it possible to detect the deformation occurring across and within plate margins by traditional geodetic surveying methods (e.g.triangulation) if observations can be repeated over several decades. In New Zealand, the first-order triangulation of the period 1925 to 1946 has been repeated in part by resurveys in the years 1976 to 1984 as part of a coordinated programme to study present-day crustal deformation (e.g. Bevin et al, 1984).

The results of such surveys can be discussed either in terms of rigid block motion, which can impose a degree of arbitrariness in the choice of blocks and boundaries, or more generally in terms of a combination of continuous and discontinuous velocity fields (e.g. Reilly 1982b, 1987). In the latter case, the intrinsic deformation of the crustal rocks can be described both in terms of simple mechanical concepts such as shear strain and dilatation, and of more complex quantities such as bending which are an expression of the variation of strain from place to place.

2 POSITION AND GRAVITY ON A DEFORMING EARTH

2.1 Rates of Change

We assume that any material point P on the earth's surface can be characterised by a three-dimensional position vector u^r, referred to some reference frame rotating with the earth, and by a scalar gravity potential W. On a deforming earth, both the position vector u^r and the potential W will change with time, and their rates of change can be denoted by

$$\hat{u}^r = \frac{\partial u^r}{\partial t} \tag{2.1}$$

$$\hat{W} = \frac{\partial W}{\partial t} \tag{2.2}$$

2.2 Deformation Tensor and the Gravity-change Vector

Both the velocity field $\hat{u}^r$ and the potential change $\hat{W}$ will vary from place to place, and this variation can be expressed by the first spatial derivative of each quantity, viz. by the deformation-rate tensor $\hat{u}^r_s$, and the gravity-rate vector $\hat{W}_r$.

2.2.1 Deformation Rate Tensor $\hat{u}^r_s$

The deformation rate tensor $\hat{u}^r_s$ at a point P expresses both the intrinsic deformation of neighbouring material (i.e.the relative motion of neighbouring particles), as well as the mean rotation of the neighbouring particles as a set with respect to the external reference frame. In particular, if p^r is the relative position vector of a point Q with repect to P, i.e.

$$p^r = u^r(Q) - u^r(P), \tag{2.3}$$

then to a first approximation the rate of change of p^r is given by

$$\hat{p}^r = \hat{u}^r_s p^s \tag{2.4}$$

where $\hat{u}^r_s$ is evaluated at the mid-point of the line PQ.

The relative position vector p^r can be decomposed into a magnitude s (length of the line PQ), and a unit vector l^r, as

$$p^r = s \, l^r \tag{2.5}$$

2.2.2 Gravity-change Vector $\hat{W}_r$

The gravity vector W_r at a point is the spatial gradient of the scalar potential W, and can be decomposed into the intensity of gravity g and the unit zenith vector v_r as

$$W_r = g \, v_r \tag{2.6}$$

The zenith vector v_r is complemented by orthogonal unit vectors in the east (λ_r) and north (μ_r) directions, where μ_r lies in the plane of the zenith vector v_r and the vector parallel to the earth's axis of rotation (C_r). The set $(\lambda_r, \mu_r, \nu_r)$ forms a reference frame at a point P for geodetic observations that depend on the direction of the vertical (e.g. zenith angles) or on the direction of the earth's rotation axis (e.g. azimuth, latitude).

Differentiation of (2.6) with respect to time leads to

$$\hat{W}_r = \hat{g}\, \nu_r + g\, \hat{\nu}_r \tag{2.7}$$

$$= \hat{g}\, \nu_r + g\, \varepsilon_{rst}\, \hat{G}^s\, \nu^t$$

where $\hat{g}$ is the rate of change of the intensity of gravity; $\hat{\nu}_r$ is the rate of change of the unit zenith vector; $\hat{G}^s$ is a rotation vector that expresses the rate of rotation of the local reference frame $(\lambda_r, \mu_r, \nu_r)$. In terms of the rate of change of the astronomical latitude $\hat{\phi}$ and longitude $\hat{\omega}$, it can be expressed as

$$\hat{G}_r = -\hat{\phi}\, \lambda_r + \hat{\omega}\, C_r \tag{2.8}$$

(see Reilly, 1985).

2.3 Continuum Models for $\hat{u}^r_s$ and $\hat{W}_r$

Within some limited region of the earth's surface, the deformation rate tensor $\hat{u}^r_s$ and the gravity-change vector $\hat{W}_r$ at a general point P can be expressed as Taylor's series expansions about a convenient origin P_o as

$$\hat{u}^r_s(P) = b^r_s + b^r_{st}\, y^t + \frac{1}{2!}\, b^r_{stu}\, y^t\, y^u + \dots \tag{2.9}$$

$$\hat{W}_r(P) = c_r + c_{rs}\, y^s + \frac{1}{2!}\, c_{rst}\, y^s\, y^t + \dots \tag{2.10}$$

where $y^t = u^t(P) - u^t(P_o)$ is the position vector of P relative to the origin P_o. The coefficients b^r_s, b^r_{st}, b^r_{stu} ..., and c_r, c_{rs}, c_{rst} ..., can be determined from the adjustment of repeated observations of geodetic networks by incorporating suitable transformations of the continuous functions (2.9) and (2.10) in the adjustment model (cf. Reilly, 1987).

2.4 Examples of Time-variable Terms for Geodetic Observations

On the assumption that rates-of-change are uniform, then the value of a geodetic observable x at time t can be deduced from its value at time t_o and its rate of change $\hat{x}$ as

$$x(t) = x(t_o) + (t - t_o)\,\hat{x} \tag{2.11}$$

As an example, we may take the case of the observation line from standpoint P to forepoint Q, represented by the relative position vector p^r (2.3). For a vector observation, as with GPS, the expression for the rate of change of this relative position vector is simply that given by (2.4) above, viz.

$$\hat{p}^r = \hat{u}^r_s\, p^s \tag{2.12}$$

For gravity-referenced observations (e.g. with a theodolite) we introduce the azimuth α and zenith distance β of the observation line PQ at P, and expand the relative position vector p^r as

$$p^r = s\, l^r \tag{2.13}$$
$$= s\,(\sin\alpha\,\sin\beta\,\lambda^r + \cos\alpha\,\sin\beta\,\mu^r + \cos\beta\,\nu^r)$$

It is also convenient to introduce some auxiliary unit vectors related to the reference frame (λ_r,μ_r,ν_r) and the unit vector l_r of the observation line, viz. the horizontal unit vector m_r normal to the (ν_r,l_r) plane

$$\sin\beta\, m_r = \varepsilon_{rst}\, l^s\, \nu^t \tag{2.14}$$

the horizontal projection h_r of l_r, and the "nadir" vector n_r, such that (m_r,h_r,ν_r), (l_r,m_r,n_r), and (λ_r,μ_r,ν_r) form right-handed orthogonal sets. We can then develop expressions for the rates of change of the length, azimuth, and zenith distance of an observation line, assuming both that the end-points are free to move relative to each other, and that the directions of the local gravity reference frames are free to move relative to the fixed global frame.

Differentiating (2.13) with respect to time, and introducing the rotation vector $\hat{G}_s$ of (2.8), and the auxiliary unit vectors, we have

$$\hat{p}^r = \hat{u}^r_s\, p^s = s\, \hat{u}^r_s\, l^s \tag{2.15}$$
$$= \hat{s}\, l^r + s\,\{\sin\beta\,\hat{\alpha}\, m^r + \hat{\beta}\, n^r + \varepsilon^{rst}\,\hat{G}_s\, l_t\}$$

whence the time derivatives of the observables distance (s), azimuth (α), and zenith distance (β) are given by

$$\hat{s}/s = \hat{u}_s^r \, l^s \, l_r$$

$$\sin\beta \, \hat{\alpha} = \hat{u}_s^r \, l^s \, m_r - \hat{G}_s \, n^s \tag{2.16}$$

$$\hat{\beta} = \hat{u}_s^r \, l^s \, n_r + \hat{G}_s \, m^s$$

Introducing the rate of change of the gravity vector $\hat{W}_r$ of (2.7), the terms involving $\hat{G}_s$ are

$$\hat{G}_s \, n^s = \frac{1}{g} \hat{W}_r \, \{\cos\beta \, m^r - \tan\varphi \, \sin\beta \, \lambda^r\} \tag{2.17}$$

$$\hat{G}_s \, m^s = -\frac{1}{g} \hat{W}_r \, h^r$$

Substitution of the expansions for $\hat{u}_s^r$ and $\hat{W}_r$ from (2.9) and (2.10) into (2.16) and (2.17) gives the differential coefficients of the unknown tensor parameters required to form the observation equations.

2.5   Adjustment of Coordinate Parameters at the Reference Epoch.

The adjustment process that yields the time-variable parameters in (2.9) and (2.10) will simultaneously yield the values of the coordinate parameters of the network stations at the reference epoch $t = t_o$, by the method of "variation of coordinates". In the simplest case in three dimensions, these parameters will comprise at each network station a triplet of geometric coordinates (e.g. geocentric Cartesian coordinates, or ellipsoidal coordinates), and the direction of the vertical, specified by the astronomic latitude φ and longitude ω.

The evaluation of the differential coefficients for an observation line PQ can be pursued in a manner analogous to that for the time-variable terms. For observations of both the length and direction of the line PQ, we require the differentials, or gradient vectors, of the scalar observed quantities (s,α,β) with respect to displacement of each of the points P and Q, and also with respect to a change in the direction of the vertical at the standpoint P. In accordance with the linearisation procedures usually adopted, we can deal with each of these variations separately.

A general expression for the gradient vectors of the observed quantities is obtained by taking the covariant derivative of (2.13) as

$$p_s^r = s_s \; l^r + s \; \{\sin\beta \; \alpha_s \; m^r + \beta_s \; n^r + \varepsilon^{rut} \; G_{us} \; l_t \} \tag{2.18}$$

in which

$$G_{us} = -\varphi_s \; \lambda_u + \omega_s \; C_u \tag{2.19}$$

is a tensor expressing the rotation of the triad of vectors $(\lambda_r, \mu_r, \nu_r)$ at P for a small displacement of P (Reilly, 1985); i.e. from (2.3)

$$p_s^r = u_s^r (Q) - u_s^r (P) \tag{2.20}$$

If we hold Q fixed, and allow only P to move, then

$$p_s^r = 0 - \delta_s^r \tag{2.21}$$

Holding the direction of the vertical also fixed (i.e. $G_{us} = 0$), then the effect on (s, α, β) of a displacement of P is given by

$$s_s \; l^r + s \; \{\sin\beta \; \alpha_s \; m^r + \beta_s \; n^r \} = - \; \delta_s^r \tag{2.22}$$

or for the individual observables

$$s_s = - l_s$$

$$\sin\beta \; \alpha_s = -\frac{1}{s} \; m_s \tag{2.23}$$

$$\beta_s = -\frac{1}{s} \; n_s$$

It is obvious from (2.20) that the effect of holding P fixed and allowing Q to move will be to reverse the signs in (2.23).

For the effect of the variation in the direction of the vertical, we now hold the points P and Q fixed, i.e. $p_s^r = 0$ in (2.18), whence

$$s_s \; l^r + s \; \{\sin\beta \; \alpha_s \; m^r + \beta_s \; n^r \} = - s \; \varepsilon^{rut} \; G_{us} \; l_t \tag{2.24}$$

The distance s is, of course, independent of the direction of the vertical, and $\varepsilon^{rut} l_t l_r = 0$ from (2.24). Evaluation of the dependence of the observed angles (α, β) with respect to changes in latitude and longitude at P is aided by introducing the set of vectors (i_r, j_r, k_r) in the directions of the coordinates (ω, φ, W), and conjugate to the gradient vectors $(\omega_r, \varphi_r, W_r)$, so that from (2.19)

$$G_{us} \, i^s \; = \; C_u \tag{2.25}$$

$$G_{us} \, j^s \; = \; -\lambda_u$$

From (2.24) and (2.25) it follows that the gradients of (α, β) with respect to a change in longitude at P are

$$\sin\beta \; \alpha_s \, i^s \; = \; -G_{us} \, i^s \, n^u \; = \; -C_u \, n^u \tag{2.26}$$

$$\beta_s \, i^s \; = \; G_{us} \, i^s \, m^u \; = \; C_u \, m^u$$

and with respect to a change in latitude at P are

$$\sin\beta \; \alpha_s \, j^s \; = \; -G_{us} \, j^s \, n^u \; = \; \lambda_u \, n^u \tag{2.27}$$

$$\beta_s \, j^s \; = \; G_{us} \, j^s \, m^u \; = \; -\lambda_u \, m^u$$

which completes the set of differentials for variation of both the positions of the end points of the line PQ, and of the vertical at P.

3   THE HORIZONTAL STRAIN FIELD

To give a simple illustration of the principles involved, we shall use an example in which the variation $\hat{W}_r$ of the gravity vector field will be ignored, and $\hat{u}^r_s$ taken to represent only the two-dimensional (horizontal) components of the deformation rate, determined from repeated triangulation measurements.

3.1 Dilatation and Shear

The two-dimensional tensor $\hat{u}^r_s$ represents both the intrinsic deformation of the horizontal surface (i.e. the relative motion of neighbouring particles), as well as the mean rotation, about a vertical axis, of the neighbouring particles as a set, with respect to the external reference frame. It can be decomposed as

$$\hat{u}_s^r = \hat{\Delta}\,\delta_s^r + \hat{\gamma}\,(j^r\,j_s - k^r\,k_s) - a^{rp}\,\varepsilon_{ps}\,\hat{\Omega} \tag{3.1}$$

in which

δ_s^r is the Kronecker tensor,

a^{rp} is the metric tensor,

ε_{ps} is Ricci's alternating tensor,

$\hat{\Delta} = \frac{1}{2}\hat{u}_r^r$ is the rate of areal dilatation,

$\hat{\Omega} = -\frac{1}{2}a^{rt}\,\varepsilon_{ts}\,\hat{u}_r^s$ is the rate of mean rotation (vorticity),

(j_r, k_r) is a pair of orthogonal unit vectors repesenting the directions of maximum and minimum relative lengthening (extension) respectively,

$\hat{\gamma} = \frac{1}{2}(j^r\,j_s - k^r\,k_s)\,\hat{u}_r^s$ is the magnitude of the shear strain rate.

From the two-dimensional analogue of (2.4) and (2.16), the rate of extension, or relative lengthening, of the line element PQ is the component of this relative velocity in the direction of the line element PQ, scaled to unit length, i.e.

$$\hat{e} = \hat{u}_s^r\,l^s\,l_r \tag{3.2}$$

while the rate of rotation of the line element PQ is given by the analogous component normal to the direction of the line element PQ, i.e.

$$\hat{r} = \hat{u}_s^r\,l^s\,m_r \tag{3.3}$$

where m_r is a unit vector orthogonal to l_r (rotated anticlockwise through one right-angle).

In terms of (3.1), the expressions for the rates of extensional strain and of rotation of a line element can be written as

$$\hat{e} = \hat{\Delta} + \hat{\gamma} \, (j^r \, j_s - k^r \, k_s) \, l^s \, l_r \qquad\qquad (3.4)$$

$$= \hat{\Delta} - \hat{\gamma} \cos 2\psi$$

$$\hat{r} = \hat{\Omega} + \hat{\gamma} \, (j^r \, j_s - k^r \, k_s) \, l^s \, m_r \qquad\qquad (3.5)$$

$$= \hat{\Omega} - \hat{\gamma} \sin 2\psi$$

where ψ is the angle between the line element l_r and the direction of maximum relative shortening k_r.

3.2 Bending

The tensor $\hat{u}_{st}^r$ has, in two dimensions, 6 independent components to describe the spatial variation of deformation at a point. One expression of this variation of deformation, or of the rate of change of shear and dilatation, for instance, is obtained by differentiating (3.1) as

$$\hat{u}_{st}^r = \hat{\Delta}_t \, \delta_s^r + \hat{\gamma}_t \, (j^r \, j_s - k^r \, k_s) - a^{rp} \, \varepsilon_{ps} \, \hat{\Omega}_t \qquad\qquad (3.6)$$

where $\hat{\Delta}_t$ and $\hat{\Omega}_t$ are the gradient vectors of the rotational invariants $\hat{\Delta}$ and $\hat{\Omega}$ respectively, and $\hat{\gamma}_t$ the gradient of the magnitude of the shear strain rate; the 6 independent components of $\hat{u}_{st}^r$ are provided by the pairs of components of each of these 3 vectors.

An alternative and more readily comprehensible expression is to introduce the concept of bending - the two-dimensional analogue of the one-dimensional bending of a beam, for example. In the case of homogeneous strain, when the tensor $\hat{u}_{st}^r$ is zero, straight lines of material particles remain straight after deformation, but in the case of heterogeneous strain, a straight line of material particles will become bent after deformation.

The rate of bending $\hat{\beta}$ of a line of particles is defined as the spatial gradient, in the direction of the line, of the rate of rotation of the line; thus by differentiating (3.3) and taking the component in the direction l^t

$$\hat{\rho} = \hat{r}_t \; l^t = \hat{u}_{st}^r \; l^s \; m_r \; l^t \tag{3.7}$$

The bending rate $\hat{\rho}$ is specified by 4 independent components, and must be complemented by some other parameter, such as the gradient of the dilatation rate, to specify completely the heterogeneous strain rate at a point. It is a cubic function of direction, and thus there will be either one or three axes of maximum bending, and one or three axes of zero bending. We shall denote, in the figures, the magnitude and direction of maximum bending by a vector symbol of length proportional to the magnitude of the bending rate, and direction normal to the axis of maximum bending and directed toward the concave side.

4   THE CONVERGENT PLATE MARGIN

4.1   The Plate Boundary in New Zealand

The boundary between the Indian and Pacific lithospheric plates passes through the New

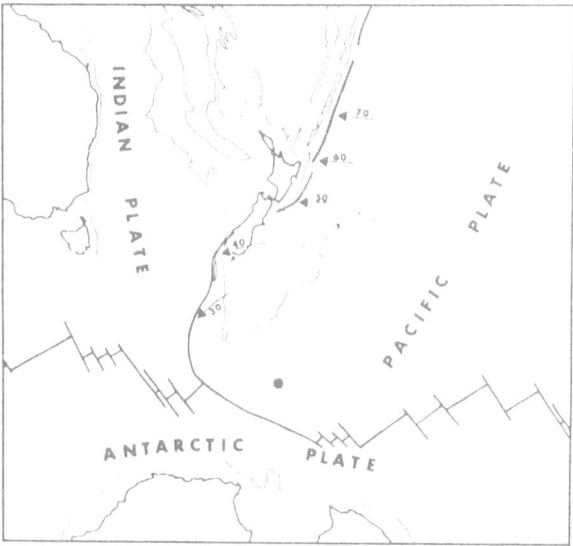

Fig 1. The three lithospheric plates near New Zealand. The inferred relative motion between the Pacific and Indian plates is equivalent to rotation about the pole shown south of New Zealand; the arrows indicate the velocity (in mm/year) of the Pacific with respect to the Indian plate. (After Walcott, 1982).

Zealand land mass (Figs. 1, 2), changing character as it does so. To the north, the Pacific Plate is subducted along the Tonga-Kermadec Trench. It continues to do so to the south, along the Hikurangi Trench, plunging obliquely westward beneath the North Island of New Zealand as far south as the Chatham Rise. Conversely, to the south the Indian Plate is subducted on the line of the Macquarie Trench and Ridge, plunging eastward beneath the Fiordland region of Southwest New Zealand. Between these two systems is a transform fault system, marked also by active mountain-building, and dominated by a major strike-slip fault, the Alpine Fault, which splays out into a series of strike-slip faults in the northern part of the South Island. Fig. 1 shows the velocity vectors of the motion of the Pacific Plate with respect to the Indian Plate. The locations of the subducting slabs are clearly demonstrated in a map of intermediate and deep-focus earthquakes (Fig. 3).

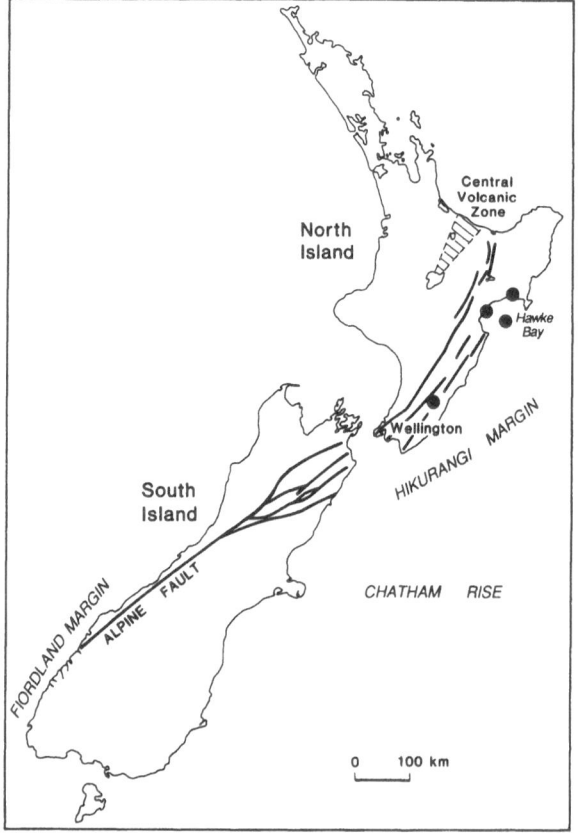

Fig 2. The plate boundary and major transcurrent fault systems in New Zealand.

The exposed land area of New Zealand, composed of fragments of the continent of Gondwana from late pre-Cambrian age onward, and Cenozoic sediments accreted since the parting from Australia, is being steadily deformed by the continuing inter-plate motion, with active seismicity, and with volcanism in the North Island in a zone of "back-arc spreading" driven by the descending Pacific Plate.

Since the motion at the plate boundary is obliquely convergent, its effects are mainly seen as dextral strike-slip faulting parallel to the boundary, and uplift (with crustal thickening) and thrust-faulting arising from the component of motion normal to the boundary.

Fig 3. Earthquakes in New Zealand in the years 1964 to 1987, with focal depths greater than 40 km, and magnitudes greater than 4.0. The symbol sizes increase in steps of 1 unit of magnitude. (After Reyners, in press).

4.2   Strain at an Obliquely Convergent Margin

It is essential to the concept of a plate boundary that there be some form of discontinuous motion between the two plates, and it will be assumed that such motion is continuing at constant velocity during the time intervals of interest (i.e. between successive geodetic surveys), unless there is strong evidence to the contrary. In other words, we adopt a working hypothesis that large-scale superficial crustal deformation is continuous in time as well as space, and that discontinuous events (faulting accompanying an earthquake) are localised in space as well as time. The occurrence of a large earthquake - large in relation to the space-time window being investigated - could invalidate this assumption.

It will be further assumed that, whatever the driving mechanism (i.e. slab-pull or ridge-push), the interaction between two converging plates takes the form of frictional slip on the surface of the descending slab, i.e. that surface tractions will be isotropic, and proportional to the relative velocity across the interface. We may then postulate that this will produce, in the

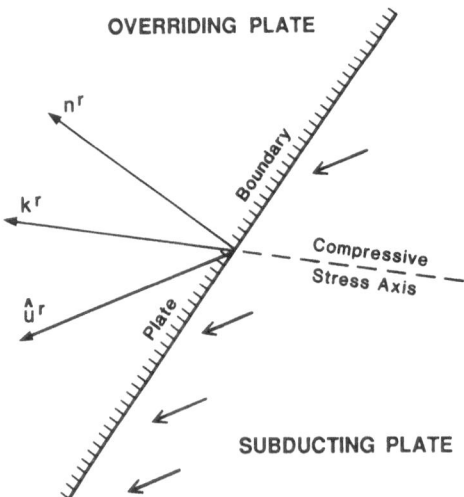

Fig 4. Stress at an obliquely convergent plate margin, where the two plates interact by frictional slip on the surface of the subducting slab. The vector n^r is the normal to the boundary; $\hat{u}^r$ the vector of velocity of the subducting slab with respect to the overriding plate, k^r is the vector in the direction of the axis of predicted maximum relative horizontal compression, taken as the bisector of the angle between n^r and $\hat{u}^r$.

hanging wall, or over-riding plate, close to the boundary, a stress field whose direction will be a function of the relative direction of plate motion and plate boundary (cf. Walcott, 1978). For relative plate motion normal to the plate boundary, we can expect the maximum principal horizontal compressive stress to be likewise normal to the boundary and parallel to the direction of motion of the descending slab. For a transform fault, with relative plate motion parallel to the plate boundary, we can expect the maximum principal horizontal compressive stress to be at 45 deg to the plate boundary, and thus to the direction of motion of the "footwall" slab. The general (oblique) case is shown in Fig. 4, where the direction of the axis of maximum principal horizontal compressive stress in the overriding plate bisects the angle between the normal to the plate boundary, and the direction of motion of the subducting slab.

This model is appropriate as a test for the deformation near the boundary of the over-riding Indian Plate on the north-east coast, or "Hikurangi Margin", of New Zealand.

5   GEODETIC SURVEYS IN NEW ZEALAND

The first-order geodetic network for the "Hikurangi Margin" of New Zealand is shown in Fig. 5 (Reilly, in press). Observations were made between about 1925 and 1946. The chains of triangulation that were repeated in the years between 1976 and 1984 are shown in the same figure; this time EDM measurements were made in addition to a repetition of the triangulation.

All observations have been reduced simultaneously in a three-dimensional adjustment (e.g. Reilly, 1982a), incorporating a continuum model following the method developed by Bibby (1973) and described further by Reilly (1982b). This model included terms corresponding to the fourth spatial derivative of a constant-velocity field. The scale, or dilatation term, was excluded, as there were no satisfactory repeated determinations of length. The solution then yielded 16 independent parameters of the model of intrinsic deformation, plus 1 parameter describing the equivalent rigid-body rotation (i.e. the coefficients of tensors of rank 2 to 4 inclusive), together with the 153 independent terms of the corresponding error covariance matrix.

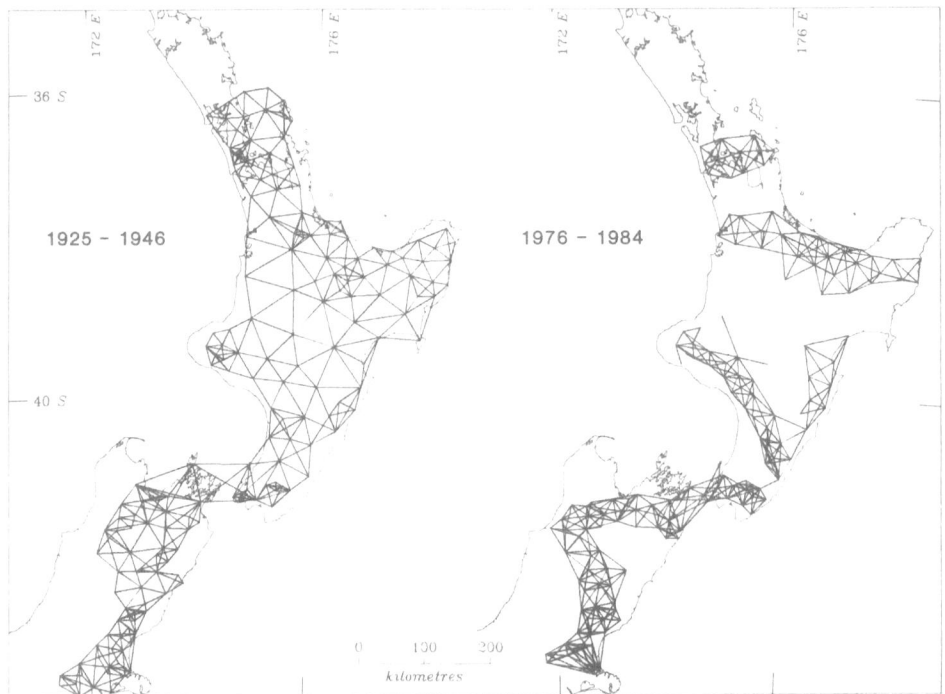

Fig 5. First order triangulation network in New Zealand, observed in the years 1925 to 1946 (left), and repeated in part with both angular and EDM measurements in the years 1976 to 1984 (right).

6 DEFORMATION ALONG THE HIKURANGI MARGIN

6.1 Results of the Analysis

A representative set of values of shear-strain rate is plotted in Fig. 6, in which each symbol is aligned in the direction of the axis of maximum relative shortening (compression), with length proportional to the magnitude of the shear strain rate. The circular sectors indicate the limits corresponding to 2 standard deviations (SD) in both magnitude and direction. Points where the result was less than 2 SD in either magnitude or direction are denoted by a small circle.

Fig. 7 shows the results of the analysis of bending. Axes of zero bending are denoted by broken lines, and the concave normals to the directions of maximum bending by full lines. Again, the sector symbols denote limits of 2 SD in both magnitude and direction; the existence

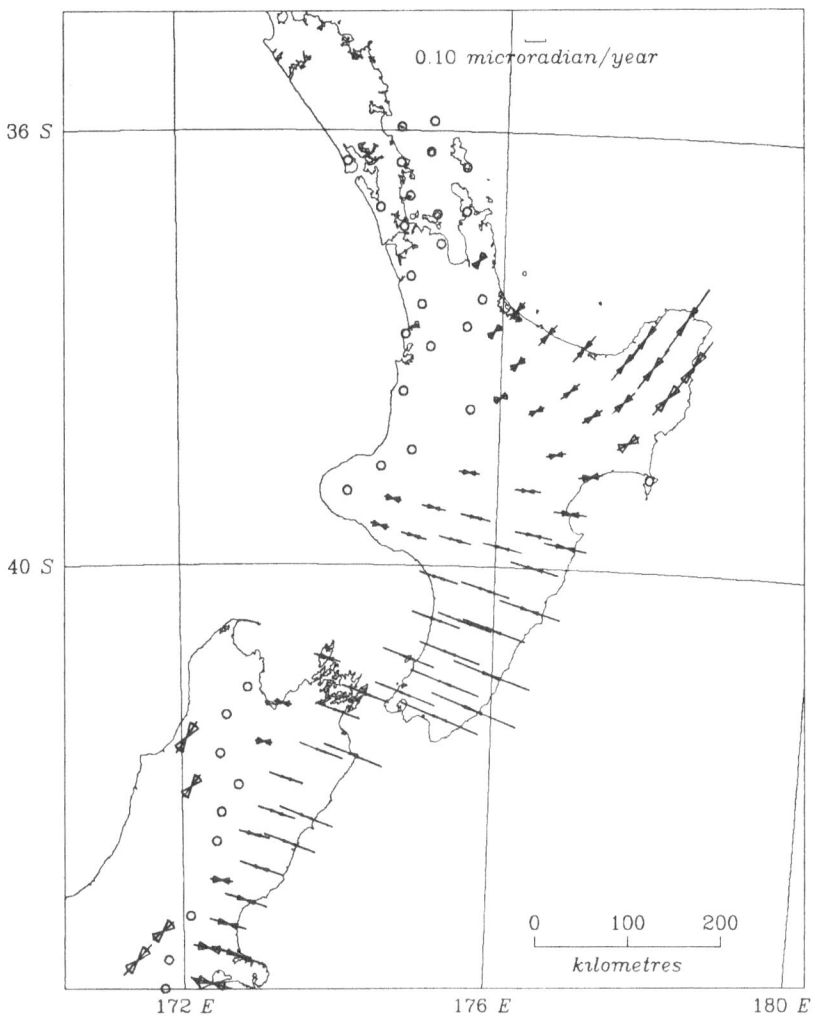

Fig 6. Horizontal shear strain rates deduced from the repeated geodetic measurements by fitting a 16-term continuum model of deformation. The symbols indicate the axis of maximum relative shortening, the length being proportional to the magnitude of the shear strain rate; the sector symbols give the limits of 2 SD in both magnitude and direction. (After Reilly, in press).

of components within these limits is denoted by a small circle. Note that while the number of axes of either zero or maximum bending is either 1 or 3, in the figure, the depiction of a number of axes was suppressed if their associated values were not significant.

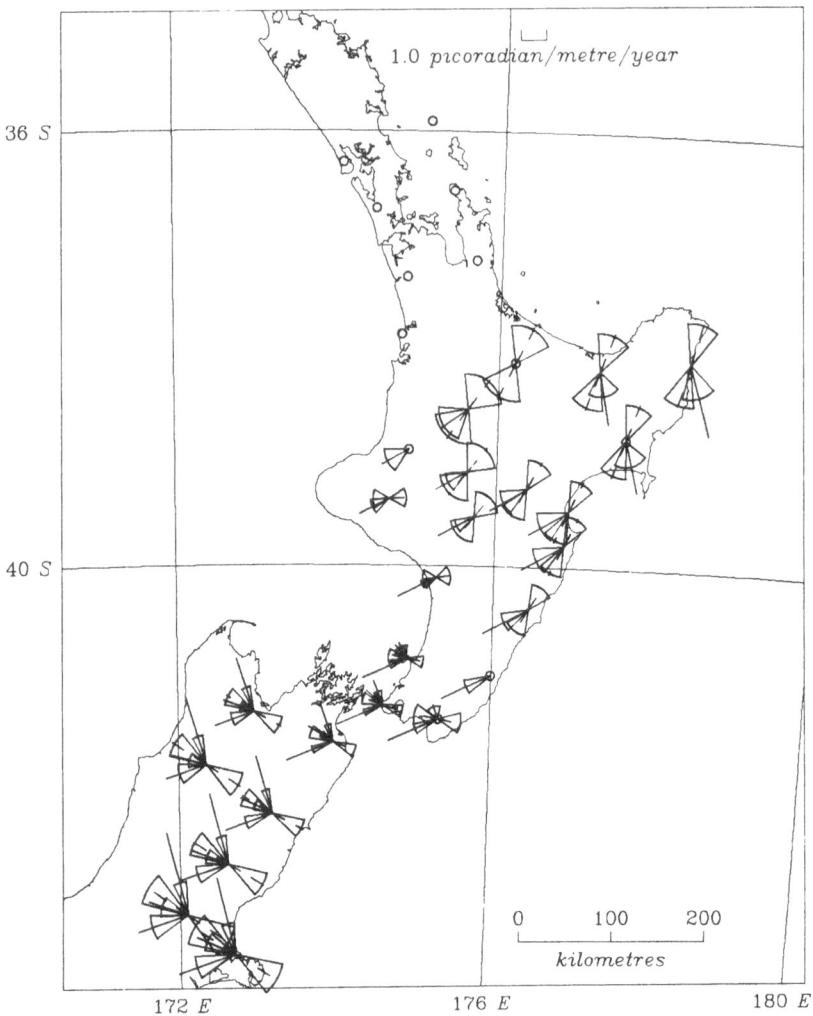

Fig 7. Horizontal bending deduced from the same solution as that of Fig 6. Broken lines represent axes of zero bending; full lines the magnitude of the bending rate, and the direction of the normal to the axis of maximum bending rate, oriented toward the concave side. Sector symbols give the limits of 2 SD in both magnitude and direction. (After Reilly, in press).

6.2   Discussion

The results displayed in Figs. 6 and 7 are computed in each case from only 16 independent parameters, so some of the apparent regularity may result from the restricted size of the approximating function. Nevertheless, the striking feature of Fig. 6 is the constant alignment of the axis of maximum shortening at about 110/290 deg, south of Hawke Bay, with the magnitude of the shear strain rate diminishing westward, away from the plate boundary. With the plate boundary at an azimuth of about 40/220 deg, (and thus 310 deg for the normal toward the hanging wall), and a relative velocity of the Pacific Plate, at the latitude of Wellington, in the direction due west, i.e. 270 deg, the predicted axis of maximum compressive stress would be (1/2)(310 + 270), or 110/290 deg, agreeing with the direction deduced from the geodetic data. This suggests that, whatever the fine structure of the deformation in terms of local faulting and folding, the deformation at the scale of the first-order geodetic network (ca. 40 km in space and 50 years in time) is that expected to be produced by continuous frictional slip on the surface of the descending Pacific Plate.

In the north-west the results suggest relative extension in the east-west direction, as found by Sissons (1979) in a study of both first and lower order triangulation in the volcanic zone. This direction is maintained further to the east, north of Hawke Bay, and its interpretation is rather ambiguous. While the Central Volcanic Zone is seen to be rifting as a result of "back-arc spreading" under the pull of the sinking slab, the east coast north of Hawke Bay is a region of rapid uplift (ca. 1-2 mm/year over the last 100 000 years), as attested by the evidence of many raised beaches and river terraces (Pillans, 1986). Offshore, an accreting sedimentary wedge shows structures attributable to compression normal to the plate boundary. Either the data here are deficient (only a single chain of repeated triangulation crosses this zone), or it is possible that trenchward gravitational slumping of the superficial layers is carrying with it the geodetic monuments on the mountain tops, for the rocks are largely incompetent mudstones of late Tertiary age.

The pattern of bending shows two significant features. The east coast north of Hawke Bay exhibits prominent bending with concavity to the south-east, which is a further sign of the apparent change to anomalous deformation in this area. In the South Island, complex bending with concave directions largely in the north-west quadrant marks the southern extremity of the subducting Pacific Plate, and the zone of transition to a transform fault regime, marked by strike-slip faulting and rapid mountain-building.

Large earthquakes occurred at two localities (Fig. 2) within the space-time window of the repeated triangulations discussed here: in Hawkes Bay the Napier Earthquake of 1931 Feb 2 (M = 7.8) which produced extensive uplift (ca. 1 m) on land, but no visible faulting, together with lesser shocks of M = 7.1 on 1931 Feb 13, and M = 7 on 1932 Sep 15; and the Wairarapa earthquake of 1942 Jun 24 (M = 7), again with no visible surface break (Smith and Berryman, 1986). None of these events appears to have caused a notable discontinuity in the pattern of horizontal deformation, though the relation between the earthquakes in Hawkes Bay and the change in strain regime there needs more investigation.

7 CONCLUSIONS

The analysis of repeated geodetic observations by incorporating a continuum model of deformation, and the analysis of the results in terms not only of rates of strain, but also of higher order terms such as bending, throws new light on the determination and description of crustal deformation, particularly where continental crust is involved in the zone of oblique convergence between two lithospheric plates.

8 ACKNOWLEDGMENTS

I wish to thank Dr H M Bibby for his critical reading of the manuscript, Miss Catherine Hourihan for preparation of the text, and Mrs Carolyn Hume for drawing the figures. Part of this work was carried out under a grant from Deutsche Forschungsgemeinschaft while I was a guest at Universitaet der Bundeswehr, Munich.

9 REFERENCES

Bevin, A.J., Otway, P.M., Wood, P.R., 1984: Geodetic monitoring of crustal deformation in New Zealand. In R.I. Walcott (Ed.) "An introduction to Recent Crustal Movements of New Zealand". Royal Society of New Zealand Miscellaneous Series 7, Wellington.

Bibby, H.M., 1973: The reduction of geodetic survey data for the detection of earth deformation. Geophysics Division Report No 84, Wellington.

Pillans, B., 1986: A late Quaternary uplift map for North Island, New Zealand. in W.I. Reilly and B.E. Harford (Eds.) "Recent Crustal Movements of the Pacific Region", Royal Society of New Zealand Bulletin 24, 409-417.

Reilly, W.I., 1982a: Three-dimensional adjustment of geodetic networks using gravity field data. Veroeff. Deutsche Geodaetische Kommission, B, 258/VII: 142-156.

Reilly, W.I., 1982b: Three-dimensional kinematics of earth deformation from geodetic observations. Veroeff. Deutsche Geodaetische Kommission, B, 258/V: 207-221.

Reilly, W.I., 1985: Differential geometry of a time-varying gravity field. Bolletino di Geodesia e Scienze Affini 44: 283-293.

Reilly, W.I., 1987: Continuum models in earth deformation analysis. in H. Pelzer and W. Niemeier (Eds.) "Determination of heights and height changes", Bonn, Duemmler, 557-569.

Reilly, W.I., (in press): Horizontal crustal deformation on the Hikurangi Margin. New Zealand Journal of Geology and Geophysics.

Reyners, M., (in press): New Zealand seismicity 1964-87: an interpretation. New Zealand Journal of Geology and Geophysics 32.

Sissons, B.A., 1979: "The horizontal kinematics of the North Island of New Zealand". Unpublished Ph.D. thesis, Victoria University of Wellington, Wellington, 118pp.

Smith, W.D., Berryman, K.R., 1986: Earthquake hazard in New Zealand: inferences from seismology and geology. in W.I. Reilly and B.E. Harford (Eds.) "Recent Crustal Movements of the Pacific Region", Royal Society of New Zealand Bulletin 24, 223-243.

Walcott, R.I., 1978: Present tectonics and Late Cenozoic evolution of New Zealand. Geophysical Journal of the Royal Astronomical Society 52: 137-164.

Walcott, R.I., 1982: The gates of stress and strain. In M.M. Cresswell (Ed.) "Large Earthquakes in New Zealand". Royal Society of New Zealand Miscellaneous Series 5, 11-16.

Four Dimensional Adjustment of the Finnish First-Order Triangulation: Results of a Test Computation

J. Kakkuri and R. Chen
Finnish Geodetic Institute
Helsinki, Finland

ABSTRACT:

Geodetic observations performed in a certain area at different periods of time provide us with an opportunity to study the local strain pattern as well as the crustal deformations possibly taking place in the area for such a study; the area in question belongs geologically to the Fennoscandian Shield.

The strain patterns found agree with those obtained from geological considerations. Because the time span between the triangulation and the trilateration measurement epochs used is short, it would be dangerous to interpret the deformations found as real without careful investigation of various error sources affecting the triangulation and trilateration data. Such an investigation will be included with the final study, which is to extend the present one to cover the whole territory of Finland.

1. INTRODUCTION

The network of the first-order triangulation of Finland can be divided into two parts: one is the triangulation chain observed around the 1930's; the other is the trilateration network filling in the holes between the triangulation chains. This network was observed around the 1970's. This opens up, at least in principle, a possibility to study whether or not local crustal deformations have occured in the country. A triangulation chain in the South-West part of Finland was chosen for an experimental computation, cf. Figs 2a and 2b.

The experimental computation performed consisted of four phases, cf. Fig. 1. The first phase consisted of collecting the data. The terrestrial first-order triangulation data were obtained from the Finnish Geodetic Institute's files (epoch I data) while the trilateration data were supplied by the National Bureau of Survey (epoch II data). The second phase, the observational adjustment, was a 3-d adjustment using the program GEONET written by Reilly (1988). The output of this phase was a set of observables to be used in the third phase, the deformation study. In this phase, the computer program written by one of us (Chen) was used. The output of this phase was a deformation model to be used in the fourth phase, the coordinate adjustment, which was 4-d adjustment with the output of the final values for X, Y, Z and the deformation parameters. In the following the phases 2, 3, and 4 will be described in detail.

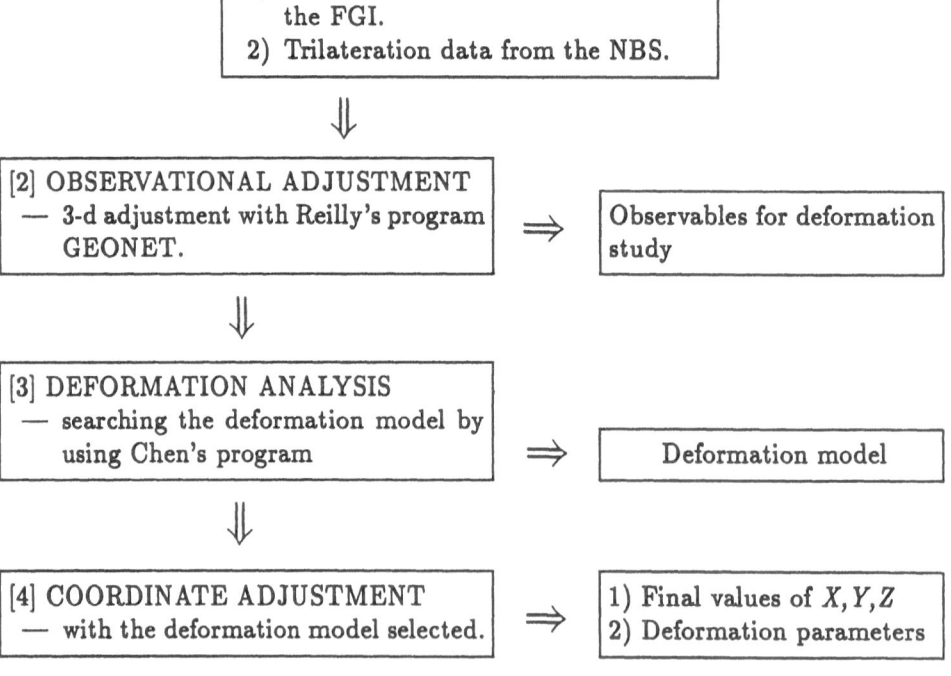

Fig.1. Different phases of the experimental computation performed.

2. OBSERVATIONAL ADJUSTMENT

2.1 Available Observations in the Area of Study

The geodetic observations available comprise of horizontal directions, EDM distances, invar baseline lengths, azimuths, and zenith distances. Also "physical" geodetic observations are available; these include astronomical longitudes and latitudes, geopotential differences from levelling and gravimetric measurements. A large number of gravimetric observations are distributed homogeneously throughout the area to be investigated; however, such measurements were not performed directly at the triangulation stations. As we mentioned above, only "geodetic" types of observations were used in our present adjustment. Zenith distance measurements were performed only at part of the stations of the test loop. This causes a configurational rank defect in the adjustment. To eliminate this defect, we use the orthometric heights of the stations at which zenith distances were not measured, to produce quasi-zenith-distances. Orthometric heights at the stations were determined by spirit levelling with an accuracy better than 0.1 m or by trigonometric levelling with an accuracy better than 0.5 m (Korhonen, 1966). The refraction corrections of the observed zenith distances were determined by a physical model given by Kakkuri (1987).

2.2 Adjustment

The general form of an observation equation for a scalar observable s in the program GEONET is:

$$s = f_1(X, Y, Z, \Lambda, \Phi, W, g) + (t - t_0)\big(f_2(b_i) + f_3(c_j)\big) + v \tag{1}$$

where f_1 is a function of the geocentric Cartesian coordinates (X, Y, Z) and of the corresponding astronomic longitude and latitude (Λ, Φ), gravity potential W, and gravity g; t is the epoch of the observation, and t_0 is the reference epoch; f_2 is a function of the parameters b_i of the strain or faulting model and f_3 a function of the parameters c_j characterising the time-varying gravity-field model; and v is the scalar residual.

Figs 2a and 2b show the configuration of the test net. In order to eliminate the datum defects from the adjustment, we fixed the Cartesian geocentric coordinates of station no. 27 as well as the observed astronomical azimuth α_{27-29}, cf. Fig. 2a. The coordinate differences obtained in the local horizontal system between the two observation epochs are shown in Table 1.

3. DEFORMATION ANALYSIS

3.1 Aspects of Crustal Deformation

The well-known post glacial uplift phenomenon taking place in Fennoscandia has been regarded by many geologists as isostatic in nature, having developed as a consequence of the release of the Earth's crust from the ice load. At the present time, its

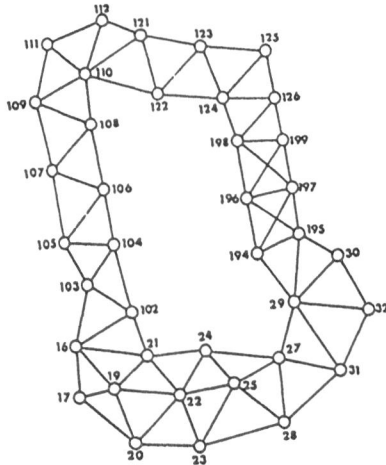

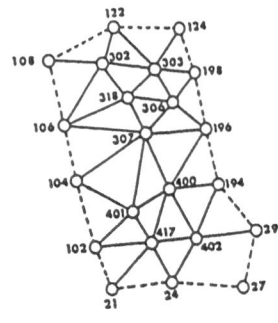

Fig. 2a. The first order triangulation chain in the test loop. The epoch of the measurement is around 1920.

Fig. 2b. The trilateration net inside the triangulation chain. The epoch of the measurement is around 1960.

Table 1. Differences between the coordinates obtained from the observational adjustments for the epochs I and II. They are given in meters according to the local horizontal coordinate system with the point no. 27 as origin.

N	D_x	D_y	D_z
21	+0.107	−0.142	+0.137
24	+0.069	−0.121	+0.116
27	0.000	0.000	0.000
29	+0.043	+0.016	+0.025
102	−0.076	−0.050	−0.188
104	+0.301	+0.057	−0.200
106	+0.402	+0.272	−0.092
108	+0.378	+0.289	+0.193
122	+0.400	+0.384	+0.255
124	+0.204	+0.477	+0.524
194	−0.048	+0.146	+0.313
196	−0.270	+0.333	+0.432
198	+0.085	+0.452	+0.519

maximum value is 1 cm/yr in the center of the area, see Fig. 3. The uplift rate in the area of our test network ranges from 0.35 cm/yr to 0.40 cm/yr. These values have been determined by repeated spirit levellings of high accuracy, cf. Kääriäinen (1966), Kakkuri and Vermeer (1985).

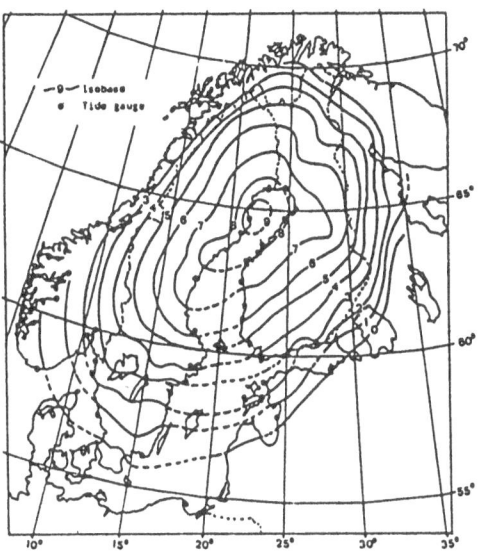

Fig. 3. Observed land uplift in the Fennoscandian area; in mm/yr.

At present, the occurrence of horizontal crustal deformation in Finland is much more uncertain than vertical deformation caused by the land uplift. Generally, Fennoscandia should be driven eastward at a rate of 1-2 cm/yr as a consequence of the spreading of the Atlantic mid-ocean ridge (cf. Artyushkov, 1983). When we observe the local area, we find different kinds of bedrock fractures with varying size everywhere in Finland. They form different mechanical discontinuities. These discontinuities include joints, fissures, faults, and large crushed zones. They divide the bedrock into more or less separate rigid blocks that together form a mosaic-like structure and originate in numerous crustal deformations. These have mostly been caused by the high horizontal compression stress which has exceeded the strength of the brittle rock. Owing to their general abundance, these fractures together form a dense network of movement planes and as a result, the whole Fennoscandian Shield is in a state resembling "plasticity" (Niini, 1987).

According to general characteristics of the bedrock fractures, we may assume a homogeneous strain deformation model.

3.2 Strain Patterns

Due to the lack of precise elevation observations, 3-d strain analysis is not possible and, therefore, only horizontal 2-d strain analysis will be discussed here.

According to the theory of homogeneous strain, the displacement vector can be given as follows:

$$u = d\mathbf{F} \cdot \mathbf{r} + u_0 \tag{2}$$

where u_0 is an arbitrary translation and

$$\mathbf{r} = \begin{pmatrix} x \\ y \end{pmatrix}$$

The displacement gradient $d\mathbf{F} = d\mathbf{R} + \mathbf{E}$ is composed of the rotation matrix:

$$d\mathbf{R} = \begin{pmatrix} 0 & \omega \\ -\omega & 0 \end{pmatrix}$$

and the symmetric strain tensor:

$$\mathbf{E} = \begin{pmatrix} e_{xx} & e_{xy} \\ e_{xy} & e_{yy} \end{pmatrix}$$

Hence it follows that equation (2) can be rewritten as:

$$
\begin{aligned}
u_x &= x \cdot e_{xx} + y \cdot e_{xy} + y \cdot \omega + c_1 \\
u_y &= x \cdot e_{xy} + y \cdot e_{yy} - x \cdot \omega + c_2
\end{aligned} \tag{3}
$$

for each point $i = 1 \cdots k$. The observations u can be adjusted under least-squares with a weight matrix $\mathbf{P_u} = \mathbf{Q_u^-}$ when a singular adjustment model is used in the observation adjustment or, alternatively, $\mathbf{P_u} = \mathbf{Q_u^{-1}}$ when a regular adjustment model is used in the observation adjustment. $\mathbf{Q_u}$ is obtained from the variance-covariance matrix of the adjusted coordinates of each epoch.

$$\mathbf{Q_u} = \mathbf{Q_x} + \mathbf{Q_{x'}} - 2\mathbf{Q_{xx'}} \tag{4}$$

$\mathbf{Q_{xx'}} = 0$ when the observations of the two epochs are independent.

Unfortunately, the unknown parameter vector

$$\mathbf{p} = [e_{xx} \ e_{xy} \ e_{yy} \ \omega \ c_1 \ c_2]^T \tag{5}$$

is related to the reference frame and, therefore, is not estimable.

On the other hand, the strain tensor $\mathbf{E}$ can be estimated directly from the repeated geodetic observations, or, in our case, from observables which are derived from the adjusted coordinates. Neither of them depends on the geodetic reference frame. E.g., Welsch (1983) has given a very detailed description about this.

It is easily shown that if the observable is a distance s between points P_i and P_j, the azimuth angle of which is A_{ij}, then the functional relationship between the linear extension e and the strain tensor components is:

$$e = e_{xx} \cos^2 A_{ij} + e_{xy} \sin 2A_{ij} + e_{yy} \sin^2 A_{ij} \tag{6}$$

and that between the distortion g of an angle between points P_i, P_j, P_k and the strain tensor components is:

$$g = e_{xy} \left(\cos 2A_{ik} - \cos 2A_{ij} \right) + \frac{1}{2} \left(e_{yy} - e_{xx} \right) \left(\sin 2A_{ik} - \sin 2A_{ij} \right) \qquad (7)$$

Linear extensions e and angle distortions g may be considered as observations $dl + v_{dl}$ and can be adjusted under least-squares with a weight matrix

$$\mathbf{P} = \mathbf{Q}_{dl}^{-1} \qquad (8)$$

where $\mathbf{Q}_{dl} = \mathbf{L}\mathbf{Q}_u\mathbf{L}^T$ in which $\mathbf{L}$ is a coefficient matrix of the function between distortion g or linear extension e and displacement vector.

For a block that contains n points, $2n - d$ independent observables can be obtained, and three parameters e_{xx}, e_{xy}, e_{yy} are expected to be solved. Here d is the datum defect.

Fig. 4 shows the divided structure of the test loop area as well as the principal strain elements in each block showing that extension takes place in southern part and northern part of the loop area, while compression occurs in its center. Movement taking place in the central block seems much more significant than that of the others; the same result can also be seen from the change of length of sides at the inner edge of the loop cf. Fig. 5. Stations no. 102 and no. 196 appear to have moved southwards. A further detailed study about this will be made in future. It could be, however, of interest to know that both of the stations mentioned are located on the vicinity of the fracture zone well known from the geological surveying of Finland (line A-A in Fig. 6).

The strain patterns we have obtained are similar to those obtained from the general geological investigation, cf. Fig. 7. Kuivamäki $et\,al.$ (1985).

3.3 Deformation model

Based on the deformation study performed, a horizontal homogeneous strain deformation model for each block is chosen here to be used in the final phase, the coordinate adjustment. Such a model has the following mathematical form:

$$u_x(x,y) = \frac{\partial u_x}{\partial x} \cdot x + \frac{\partial u_x}{\partial y} \cdot y$$
$$u_y(x,y) = \frac{\partial u_y}{\partial x} \cdot x + \frac{\partial u_y}{\partial y} \cdot y \qquad (9)$$
$$u_z(t) = (t - t_0) \cdot \delta_v$$

or more simply

$$\mathbf{u} = \mathbf{G} \cdot \delta \qquad (10)$$

where $\mathbf{u}$ is the displacement vector, δ is a vector of deformation parameters, $\mathbf{G}$ is a coefficient matrix, and x, y, z are the topocentric rectangular coordinates in the local horizontal system; δ_v describes the rate of land uplift in the area investigated.

To obtain the topocentric rectangular coordinates x, y, z from the geocentric Cartesian coordinates X, Y, Z, which are obtained from the observational adjustment, the following transformation is to be made:

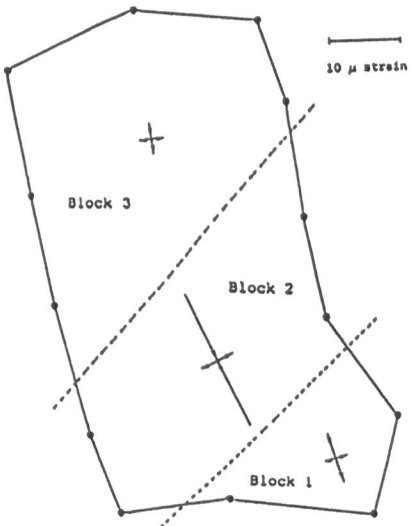

Fig. 4. Principal strain elements at each block.

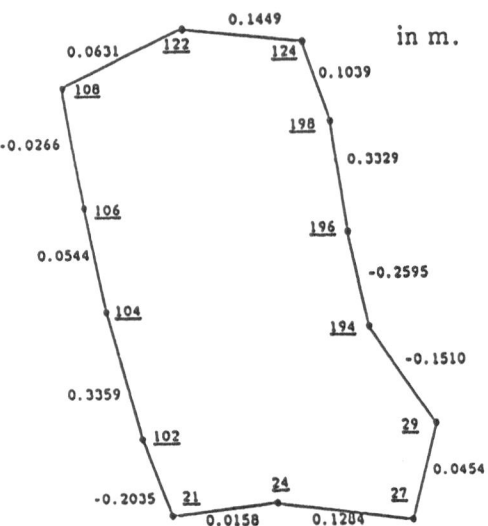

Fig. 5. Changes of the sides at the inner edge of the test loop.

Fig. 6. Fracture distribution in Finland.

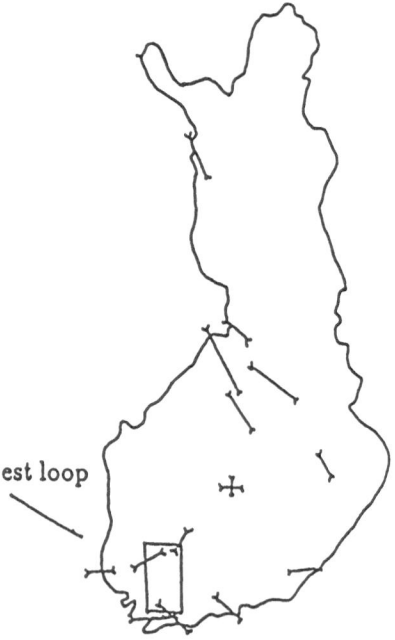

Fig. 7. Main horizontal tension direction in Finland.

$$\begin{pmatrix} x \\ y \\ z \end{pmatrix} = \mathbf{R} \cdot \begin{pmatrix} X_j - X_i \\ Y_j - Y_i \\ Z_j - Z_i \end{pmatrix} \tag{11}$$

where the rotation matrix $\mathbf{R}$ has the orthogonality property $\mathbf{R}^T = \mathbf{R}^{-1}$, and $\mathbf{R}$ itself is defined by

$$\mathbf{R} = \begin{pmatrix} -\sin\Phi\cos\Lambda & -\sin\Phi\sin\Lambda & \cos\Phi \\ -\sin\Lambda & \cos\Lambda & 0 \\ \cos\Phi\cos\Lambda & \cos\Phi\sin\Lambda & \sin\Phi \end{pmatrix} \tag{12}$$

where Λ, Φ are the astronomic longitude and latitude at the standpoint.

Similarly, the variance-covariance matrix should be written as:

$$\mathbf{Q}_{zz} = \mathbf{R} \cdot \mathbf{Q}_{XX} \cdot \mathbf{R}^T \tag{13}$$

4. COORDINATE ADJUSTMENT

After the analysis of the strain pattern, the whole loop area was divided into three blocks and in each block a horizontal homogeneous strain deformation model was considered to be valid. In this way, the mathematical expression can be written as follows:

$$\begin{pmatrix} \mathbf{r} \\ \mathbf{r}' \end{pmatrix} = \begin{pmatrix} \mathbf{I} & 0 & 0 & 0 & 0 \\ \mathbf{I} & \mathbf{B} & \mathbf{C} & \mathbf{D} & \mathbf{E} \end{pmatrix} \cdot \begin{pmatrix} \xi \\ \delta_1 \\ \delta_2 \\ \delta_3 \\ \delta_v \end{pmatrix} - \begin{pmatrix} 1 \\ 1' \end{pmatrix} \tag{14}$$

where

$$\begin{aligned} \mathbf{r} &= [\mathbf{r}_1 \cdots \mathbf{r}_n]^T & \mathbf{r}_i &= [x_i \ y_i \ z_i]^T \\ \mathbf{r}' &= [\mathbf{r}'_1 \cdots \mathbf{r}'_n]^T & \mathbf{r}'_i &= [x'_i \ y'_i \ z'_i]^T \end{aligned}$$

$\mathbf{B}, \mathbf{C}$, and $\mathbf{D}$ are the coefficient matrices of the blocks, $\mathbf{E}$ and δ_v describe the land uplift in the area investigated; δ_i is the deformation parameter vector of block i and, ξ is an unknown coordinate vector. For example $\mathbf{B}$ can be expressed as:

$$\mathbf{B} = \begin{pmatrix} \mathbf{B}_1 \\ \vdots \\ \mathbf{B}_n \end{pmatrix} \qquad \mathbf{B}_i = \begin{pmatrix} x_i & y_i & 0 & 0 \\ 0 & 0 & x_i & y_i \\ 0 & 0 & 0 & 0 \end{pmatrix} \tag{15}$$

$\mathbf{B}_i = 0$ when point i is not considered in the corresponding block. $\mathbf{C}$ as well as $\mathbf{D}$ have the same mathematical form as $\mathbf{B}$. Finally, the matrix $\mathbf{E}$ is as follows:

$$\mathbf{E} = \begin{pmatrix} \mathbf{E}_1 \\ \vdots \\ \mathbf{E}_n \end{pmatrix} \qquad \mathbf{E}_i = \begin{pmatrix} 0 \\ 0 \\ t - t_0 \end{pmatrix} \tag{16}$$

Because the rate of the land uplift is known, it is unnecessary to consider it as an unknown parameter, but it must be used to calculate the constant terms. Accordingly, the observation equations can be rewritten as follows:

$$\begin{pmatrix} v \\ v' \end{pmatrix} = \begin{pmatrix} I & 0 & 0 & 0 \\ I & B & C & D \end{pmatrix} \cdot \begin{pmatrix} \xi \\ \delta_1 \\ \delta_2 \\ \delta_3 \end{pmatrix} - \begin{pmatrix} 1 \\ 1 + E \cdot \delta_v \end{pmatrix} \tag{17}$$

or more simply

$$v = A \cdot \xi + G \cdot \delta - 1 \tag{18}$$

with a weight matrix

$$P = \begin{pmatrix} N & 0 \\ 0 & N' \end{pmatrix} \tag{19}$$

where N and N' are taken from the observational adjustment of v and v', respectively; they correspond to the points used in the last phase, coordinate adjustment; such points have been observed at both epochs.

Once we obtain the observation equations and the weight matrix, a least-squares solution can be obtained easily.

Table 2. Deformation parameters

	$\partial u_x / \partial x$	$\partial u_x / \partial y$	$\partial u_y / \partial x$	$\partial u_y / \partial y$
block 1	$+1.911 \pm 2.07$	-1.470 ± 3.31	-0.226 ± 3.30	$+2.956 \pm 3.36$
block 2	-2.421 ± 1.84	-0.661 ± 2.47	$+4.016 \pm 2.45$	$+1.851 \pm 1.79$
block 3	$+0.665 \pm 1.95$	-3.337 ± 3.19	$+3.888 \pm 2.38$	$+1.469 \pm 3.01$

The final result of adjustment is shown in Table 2. Here δ_v was taken to be zero in the coordinate adjustment.

Strain tensors obtained from the coordinate adjustment differ from those obtained from the adjustment of the observations of linear extensions, (compare Fig. 4 and Table 2), because the deformation model used in coordinate adjustment does not describe the strain pattern of each block itself, but rather the homogeneous deformation of a block which contains the former block itself plus the origin point of the coordinate system. The strain tensors of block 1 obtained from both adjustments are the same, because it contains the origin point, i.e. the station no. 27. If only the strain patterns are needed to be shown, a satisfactory result can be obtained by using the method given by Welsch (1983). To obtain a set of relative displacements, a deformation model such as that employed in our adjustment seems necessary. Analysis of the strain patterns gives a guidance to choose the deformation parameters, and deformation parameters show us a view of the relative displacements.

5. CONCLUSION

This paper reports our preliminary study on the local crustal strain pattern and deformation determined by geodetic observations. The whole test loop studied is divided into three blocks, and the principal strain elements of each block are given. Extension occurs in the northern and southern part of the test loop, while compression takes place in the central part. With the aid of the analysis of strain patterns, adjusted coordinates and a set of deformation parameters were obtained through coordinate adjustment in the area studied. It is, however, dangerous to interpret the motions found as a real crustal phenomena without making a careful investigation into the errors that affect the triangulation as well as trilateration observations.

6. ACKNOWLEDGEMENTS

The authors wish to express their indebtedness to Dr. W. Ian Reilly for giving unselfishly his program system GEONET to us for this study as well as to Dr. Martin Vermeer who has revised the English language.

7. REFERENCES

Artyushkov E. V. (1983). Geodynamics. Developments in Geotectonics 18. Elsevier.

Kakkuri J. (1987). A physical model developed for computing refraction coefficients. Rudolf Sigl. Technische Universität München, pp. 132-135.

Kakkuri J., Vermeer M. (1985). The study of land uplift using the third precise levelling of Finland. Reports of the Finnish Geodetic Institute, No. 85:1, Helsinki.

Korhonen J. (1966). Horizontal angles in the first order triangulation of Finland in 1920-1962. Publications of the Finnish Geodetic Institute, No. 62, Helsinki.

Kuivamäki A., Vuorela P. (1985). Käytetyn ydinpolttoaineen loppusijoitukseen vaikuttavat geologiset ilmiöt Suomen kallioperässä (in finnish). Report YST, vol 47. Geological Survey of Finland, Helsinki.

Kääriäinen E. (1966). The second levelling of Finland. Publications of the Finnish Geodetic Institute, No. 61, Helsinki.

Niini H. (1987). Bedrock fractures affecting land uplift in Finland. In: Perttunen M. (ed), Fennoscandian land uplift. Geological survey of Finland, Helsinki, pp. 51-54.

Reilly W. Ian (1988). A user's guide to GEONET. Operational Geodesy Software Packages. SCHRIFTENREIHE des Studiengangs Vermessungswesen der Universität der Bundeswehr München, pp253-323.

Welsch W. M. (1983). Finite element analysis of strain patterns from Geodetic observations across a plate margin. In: Vyskočil P., Wassef A.M., Green R. (eds), Recent crustal movements 1982. Developments in Geotectonics 20, pp. 57-71. Elsevier.

Precise Geodynamic Measurements in South America

E. Groten
Institute of Physical Geodesy
Technische Hochschule Darmstadt
F.R.G.

ABSTRACT:

First high precision gravity measurements carried out in 1984 were repeated in November 1987 when in a wider frame, ranging from Santa Cruz de la Sierra (Bolivia) down to Santiago de Chile and Mendoza (Argentina), a regional densified network in Northern Chile was observed. The carefully monumented regional network extends from the earthquake-active coastal area in Chile up to Salta in Argentina. The repeated measurements are considered as a first step in a longtime study where geometric vertical control will be provided by GPS-measurements. Additional geodynamic information is provided by parallel seismic and other observations. Special interest arose from the fact that briefly after the first observations in 1984 significant earthquake deformation occurred in the area of Mendoza and Santiago de Chile. As far as gravimetry is concerned, all possible error sources are being carefully considered where also absolute measurements in view of scaling errors are planned. Reference is being made with respect to those areas which appear to be decoupled from the well known uplift of the High Andes. A detailed discussion and analysis of gravimetric data is presented. Correlation with geodynamic phenomena is studied. Future prospects of the general concept "GPS-gravimetry" as a geodynamic tool for studying vertical phenomena are interpreted.

1. INTRODUCTION AND DESCRIPTION OF THE PROBLEM:

Geodetic measurements of various types became an efficient tool in studying recent crustal movements. Whereas SLR (Katsambalos et al., 1988) and VLBI observations, in combination with absolute gravimetry, serve basically the determination of large-scale tectonics (Richards et al., 1988) GPS-observations in combination with relative gravimetry (Becker et al., 1987) may serve small-scale geodynamic studies. As the high harmonics of the earth gravity field are generated within the lithosphere it is not premature to assume that with increasing accuracy we may soon need to express gravity harmonics within a mantle-fixed reference frame in form of time-dependent harmonic coefficients as we are used to do in geomagnetism (Allredge, 1988). Associated local gravity variations could be better detected in the future by gravity gradiometry (Ager et al., 1982; Brezowski et al., 1988) than in the past. The efficiency of local gravimetry depends, however, strongly in the future on a final answer to the current descussion of the "fifth force" (Schwarzschild, 1988; Groten, 1988a). An additional tool which has not yet been sufficiently exploited are long-time series of sea level (Ekman, 1988a). The postglacial uplift is a rather particular phenomenon in Scandinavia but a variety of arguments in (Ekman, 1988) can be usefully applied in general.

In a series of papers such as (Lagios et al., 1988, Groten, 1988, Groten et al., 1988, Becker et al., 1988) the implications of geotectonics and geodynamics in view of increased seismicity (Hugenschmidt, 1988), seismic risk (Jachens et al., 1983) etc. were discussed. Forecasting earthquakes is still an unresolved problem (Kawabe, 1988, Friedmann et al., 1988). Classical geodetic measurements (Archbold, 1988) besides tiltmeter, groundwater level and gravimetric measurements can be valuable tools besides space techniques. Gravimetric local results in studying geodynamics along a well studied (Roth, 1988) part of the Northern Anatolian fault was discussed in (Groten, 1988) where also the combined approach of using tiltmeters, GPS, repeated levelling and gravity together in an area of a large artificial dam in Norway was outlined. This attempt in Norway is strongly corroborated by the results found by Jachens et al. (1983). In this paper we focus on new results obtained in Southern America from repeated regional relative gravimetry in combination with GPS-observations.

Chile and Argentina are quite fascinating along their plate boundary zones (Baker et al., 1988) but northern Chile is one of the most interesting areas from a geophysical viewpoint. The areas (western part) of isostatic equilibrium and those (eastern areas) where deviations may be significant are treated in (Kono, 1988). Plate subduction in the Bolivian part of our area under investigation was discussed in (Henry and Pollack, 1988). The earthquakes of 1985/86 in the region of Valparaiso and Santiago as well as around Mendoza, Argentina, not very far from the argentine border as well as the vulcano eruptions at the end of 1988, 700 km south of Santiago de Chile, prove the existence of present seismic activity clearly. Incidentally, these earthquakes which lead to significant damage occured briefly after our first reference measurements in Santiago and Mendoza; insofar our measurements were "in due time"; the repeated gravity observations in 1987/88 may indicate geodynamic consequences of the earthquake-related deformations of the earth's crust. This area of investigation was selected by us in view of "seismic gap" and similar considerations. The first idea of substantial uplift in the coastal zone, however, did not turn out to be correct. This is seen by inspecting our gravity as well as the other recent (geomorpholgical, seismic etc.) results. The measurements in that zone need further investigations and gravity, for reasons of uniqueness of data interpretation, has to be supplemented by geometrical data which has been done now in terms of repeated GPS-observations. With present accuracies of a few microgals (if relative gravimetry is supplemented by two absolute stations at Salta and Mendoza and associated earth tidal measurements as well earlier discussed by us) in gravimetry and a few centimeters in height from GPS the approach makes sense in the now designed form.

It also has to be seen as one component of a world-wide concept where post-galacial uplift was studied by us (with interesting results) in Scandinavia, natural effects in combination with man-made effects in the Rhinegraben for an interval of about 20 years, seismicity and uplift in combination with the construction of big dams and artificial lakes was successfully studied in Norway, the gravity effects along a plate margin in Norhtern Anatolia where shear prevails was studied in Turkey. In all cases we used combination approaches in order to solve the problem of non-uniqueness in repeated gravity data.

There do exist a variety of similar attempts (Hagiwara et al., 1985) to use gravity and other geodetic measurements for earthquake prediction as well as for controlling and surveying uplift and subsidence which might be an indicator of pre-, post and coseismic motion and is of particular interest whenever abrupt or other significant variations occur.

It was interesting to realize that in our cases where anelasticity clearly did not play any role the deformations predicted from elastic modeling of loading were definitely smaller than those determined from gravimetry and other (geometric) techniques. Therefore, observations are necessary and represent a significant source of information in spite of uncertainties in related corrections and reductions which are necessary in order to correct for atmospheric and similar perturbing effects.

The earthquake prediction problem and the question of pre-, co- and postseismic deformations still deserves further investigation (Gupta, 1988), mainly because none of the existing observation techniques can alone solve the problems. As GPS and similar space methods can, together with relative gravimetry, be often and repeatedly applied, a fast and inexpensive, substantial improvement of the situation can be expected from them.

When we compare gravimetry with satellite techniques we should realize the fundamental difference between most geodetic measurements, on one side, where measurements are carried out between a discrete sender and a discrete (pointwise) antenna or target and gravimetry, on the other side, where, in the proper sense of remote sensing, the integrated effect of the total environmental or surrounding masses on the sensor is measured. In case of discrete measurements we always have an "error of representation" being the difference between the discrete antenna and the surroundings. Almost nobody is interested in the behavior, displacement etc. of the target or antenna in geodynamics; rather the behavior etc. of a representative sample of the area under investigation is of interest. Therefore, gravimetry and discrete satellite techniques, such as VLBI, SLR etc., supplement each other in an ideal way; gravimetry, however, is also affected by masses and accelerations which are out of interest in many cases such as varying water masses (consequences of rainfall etc.), atmospheric density variations etc. Therefore, appropriate corrections are necessary. This problem has to seen in view of more complex signals in case of absolute gravimetry and, therefore, relative gravimetry has its specific advantages in this context.

The extent to which non-unique gravimetry is supplemented by geometric information was discussed in detail by Arnold, Heck and others in various papers.

2. NON-NEWTONIAN ASPECTS:

Even though present experiments on a "fifth" and possibly a "sixth force" are still totally inconclusive their impact on modern gravity, if these effects really exist, would be significant (Rubincam et al., 1988). There are basically two different aspects: (1) Even though in the determination of the gravity field by space techniques the gravitational constant, "big G", enters via $\mu = GM$, as determined from Kepler's third law (with M being the mass of the earth), the high harmonics are presently often based on terrain models where terrestrial density estimates enter in combination with G; (2) also analytical downward as well as upward continuation is affected by non-Newtonian effects if they actually exist; the US Air Force experiments in free space are solutions of the Laplace equation, using a high tower; the controversial borehole experiment in Greenland is a typical example of downward continuation experiments, besides various other investigations of that type, which are basically involving the Poisson equation where the problem enters via uncertainties of underground density.

There are numerous other cases where, in modern gravimetry, the duality of gravity based on μ, on one side, and directly based on G, on the other side, enter combination solutions. Therefore, even though relative gravimetry is today based on standard and calibration values set up by absolute gravimetry, which are basically independent of the numerical values adopted for G, there is a variety of ways in which G enters more or less indirectly.

As far as the relation between free air anomalies on heights is concerned, the related regression coefficient is directly proportional to density. In our joint campaigns of GPS and relative gravimetry the accuracy is not yet good enough for studying this relationship in detail. But again this is an example where density itself affects also the geodynamics use of relative gravimetry as applied to gravity variations with time. It is expected that the pioneering research by Stacey, Eckardt and others will also be relevant to this type of research in the future. We might even state that Tuck's big balance experiment results in combination with G. Mueller's recent reservoir results obtained in the Black Forest (G. Mueller, priv. comm., 1988) clearly corroborate the Newtonian law.

Recent results, such as in (Speake et al., 1988, Stubbs et al., 1989) indicate that the dependence of G on composition of material is more doubtful than earlier assumed. Consequently, limits for $G = G(r)$ derived from composition dependence for amplitude and phase in Yukawa's law can, to some extent, no longer be considered as valid. Insofar the dependence of G on distance is still more an open question than in the recent past. Results such as those by Romaides et al. (1989) must be seen from this viewpoint. Consequently, it is still unclear whether or not terrain corrections in gravimetry will be affected by errors of > 1 mgal caused by possible non-Newtonian effects. Terrain corrections etc. are of interest when Bouguer and isostatic gravity is used for detecting uplift and subsidence zones as in (Groten et al., 1988).

3. NUMERICAL RESULTS:

The comparison of gravity measurements carried out in 1985 and 1987/88 indicates accuracies of about 10 microgals for the area shown in Fig. 1. A comparison with GPS-results is not yet possible but is expected to be available in 1990 along the chain seen in this figure. With a 3σ-limit of ~ 30 microgals it is realized that maximum differences of gravity occur in those areas where earthquake activities were significant in 1984/85. However, these differences appear to be relatively large so that there may be instrumental effects in combination with gravity differences caused by geophysical reasons. On the other hand, co-seismic gravity changes and associated motion of this order of magnitude also occured in other seismic areas such as Northern India. We know that earthquake displacements are of the order of several decimeters, vertical and horizontal, for those magnitudes; but part of it can be reversible; a decimeter may correspond to 10 to 20 microgals. Unfortunately, no repeated height control was available.

Fig. 2 and 3 show the gravity changes for 1984 - 1987/88; for the argentine part only a zero measurement was available so that no differences for that interval could be evaluated. The results given in these figures represent preliminary data which need to be corroborated by further repeated gravity measurements. Moreover, qualitative control is possible from future GPS-results as demonstrated by Jachens et al. (1983) for classical techniques.

4. CONCLUSIONS:

The usefulness of combined modern geodetic measurement approaches is shown. The study of recent crustal movements and associated gravity changes, also over larger distances and under unfavorable logistic conditions, is demonstrated. But this is only a first step. More sophisticated and careful approaches are necessary in order to fully exploit the possibilities of modern geodetic techniques.

A B C – PROFILES

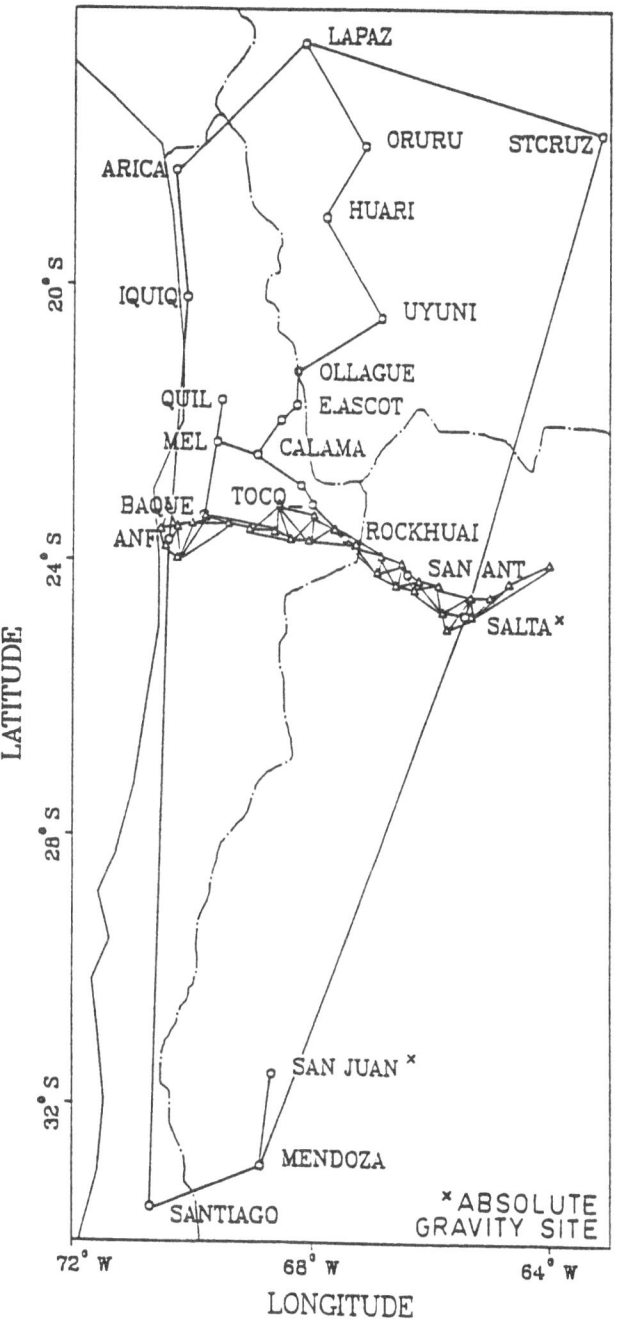

Fig. 1

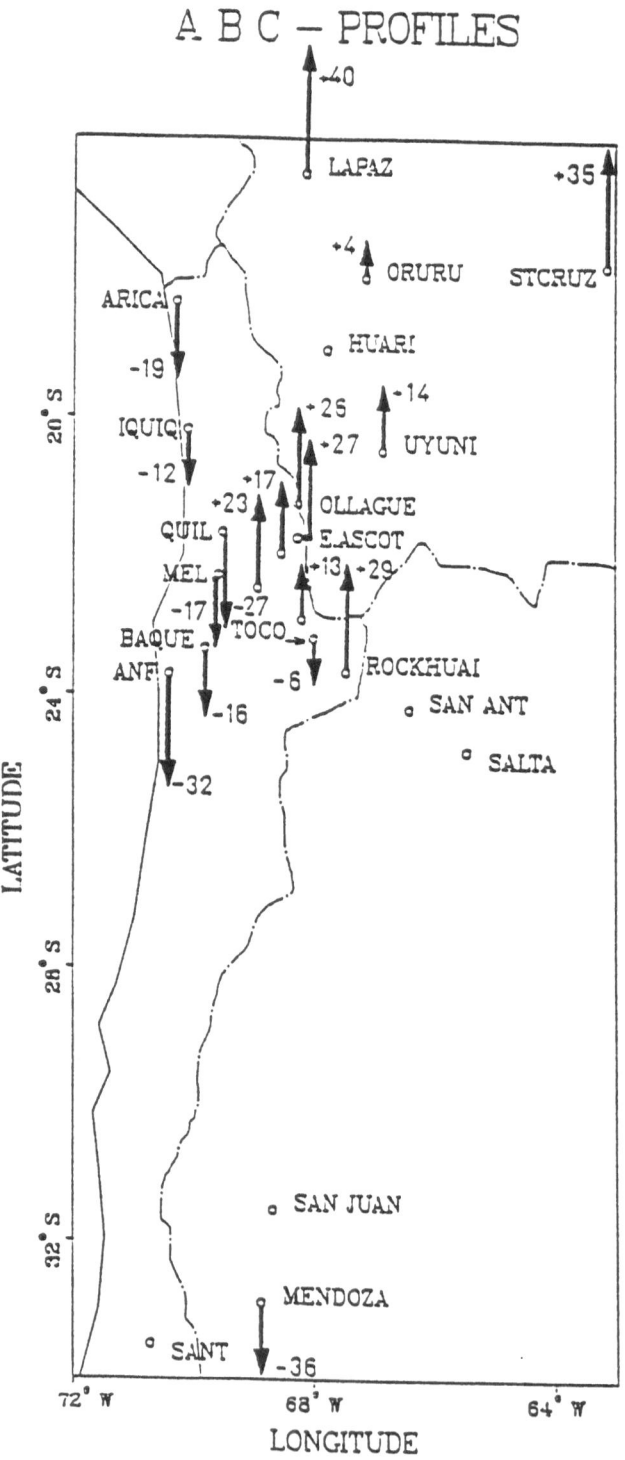

Fig. 2

Fig. 3

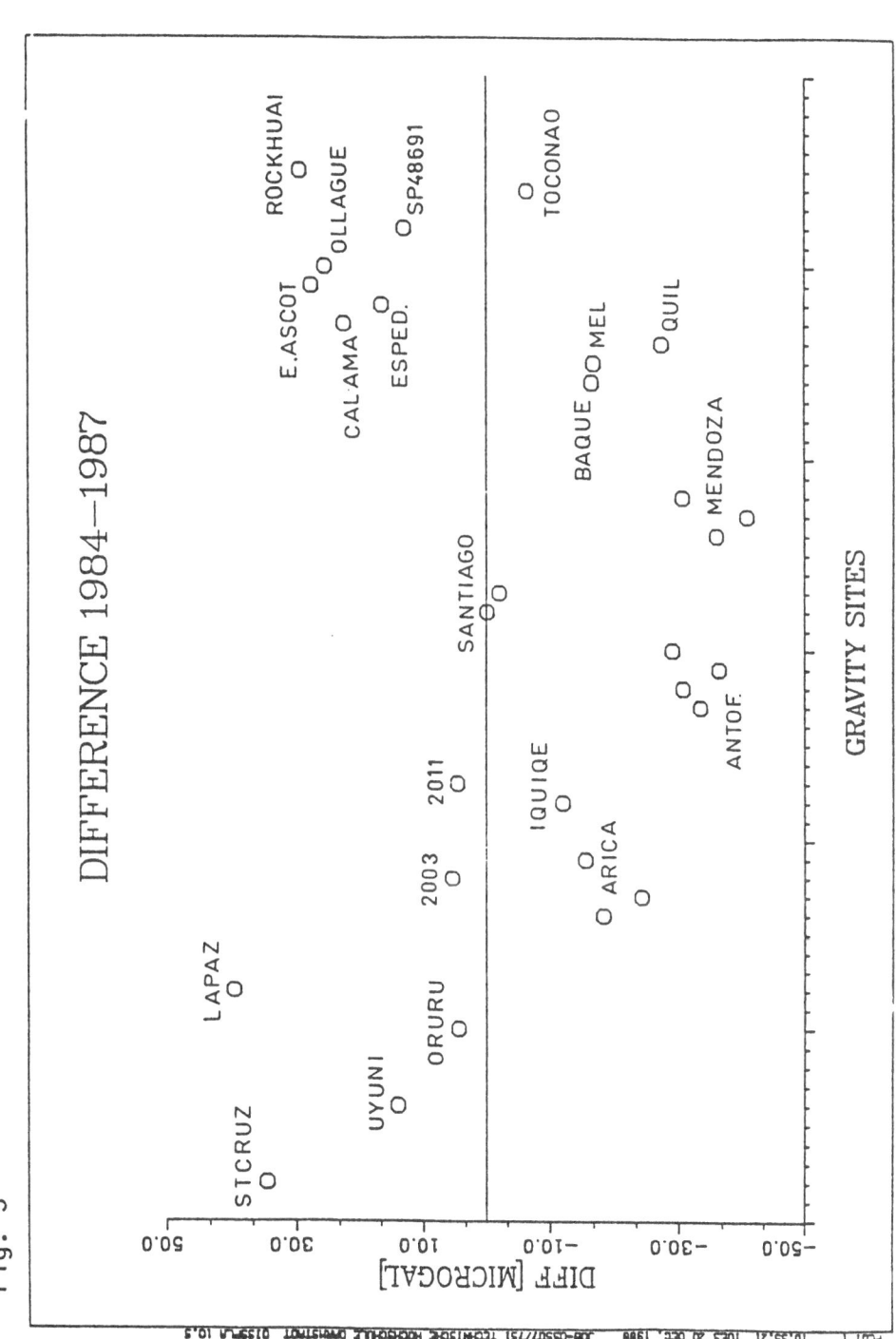

DIFFERENCE 1984−1987

ACKNOWLEDGEMENT: This paper is devoted to the memory of Prof. Ron Mather, former president of IAG-Section V. He was one of the pioneers of geodynamics. The geodetic community owes quite a lot of new ideas, brilliant concepts and impressive results to him. Many basic thoughts which are now commonly accepted by the geodetic community and which arose during his time have to be credited to him. I personally owe a variety of inspiring ideas and elucidating talks to Ron.

REFERENCES

Archbold, M.J., Ch.O. McKee, B. Talai, J. Mori and P. de Saint Ours: Electronic distance measuring network monitoring during the rabaul seismicity/deformational crisis of 1983-1985, J.G.R. 93, B10, 12123-12136, 1988

Ager, Ch.A. and J.O. Liard: Vertical gravity gradient surveys: Field results and interpretations in British Columbia, Canada, Geophysics 47, 6, 919-925, 1982

Alldredge, L.R.: New trend-trigonometric model for interpolation and prediction of the geomagnetic field utilizing the new DGRF models, J. Geomag. Geoelectr. 40, 749-759, 1988

Baker, P.F., L.A. Lawver: South American-Antarctic plate motion over the past 50 Myr, and the evolution of the South American-Antarctic ridge, Geophysical Journal 94, 377-386, 1988

Becker, M., E. Groten, A. Lambert, J.O. Liard and S. Nakai: An intercomparison of LaCoste and Romberg Model-D gravimeters: results of the International D-meter Campaign 1983, Geophys. J. R. astr. Soc. 89, 499-526, 1987

Becker, M., M. Araneda, E. Groten, O. Hirsch, T. Knöll and S. Ostrau: First results of repeated gravity measurements in the Southern Central Andes, paper presented at DFG-conference of South America, Hannover, Nov. 1988

Brezowski, S.J. and W.G. Heller: Gravity gradiometer survey errors, Geophysics 53, 10, 1355-1361, 1988

Ekman, M.: The impact of geodynamic phenomena on systems for height and gravity, Tekniska skrifter - Professional Papers, Nordic Geodetic Commission, 2nd Research School, Ebeltoft, Denmark, 1988

Ekman, M.: The world's longest continued series of sea level observations, Pageoph. 127, 1, 73-77, 1988a

Friedmann, H., K. Aric, R. Gutdeutsch, C.-Y. King, C. Altay and H. Sav: Tectonophysics 152, 209-214, 1988

Groten, E.: Gravimetric crustal motion and movement monitoring: Problems and interpretation, Journal of Geodynamics 9, 121-133, 1988

Groten, E.: Simulations for studying the Yukawa-term, ZfV 8, 366-373, 1988a

Groten, E. and M. Becker: The contribution of determinations of gravity field variations to geodynamics – A comparison with other techniques, paper presented at IAG-Symposium, Potsdam, Aug. 1988

Gupta, H.K.: Medium term earthquake prediction, EOS 69, 49, 1620-1630, 1988

Hagiwara, Y., G. Tajima, S. Izutuya, K. Nagasawa, I. Murata, S. Okubo and T. Endo: Gravity change in the Izu Peninsula in the last Decade, Journal of the Geodetic Society of Japan 31, 2, 220-235, 1985

Henry, S.G. and H.N. Pollack: Terrestrial heat flow above the Andean subduction zone in Bolivia and Peru, JGR 93, B12, 15,153-15,162, 1988

Hugenschmidt, J.: Berechnungen zur Auswirkung des Stausees Blasjo auf Neigungsmessungen. Dipl.-Arbeit, FU Berlin, FB Geowissenschaften, 1988

Jachens, R.C., W. Thatcher, C.W. Roberts, R.S. Stein: Correlations of changes in gravity, elevation, and strain in Southern California, Science 219, 1215-1217, 1983

Kawabe, I., I. Ohno and S. Nadano: Groundwater flow records indicating earthquake occurrence and induced earth's free oscillations, Geophys. Research Letters 15, 11, 1235-1238, 1988

Katsambalos, K., E. Livieratos, M. Marsella, S. Zerbini: Geometric analysis of the crustal dynamics oriented european and mediterranean network, Dept. of Geodesy and Surveying, University of Thessaloniki, Greece, 1988

Kono, M.: Mountain building in the Central Andes, abstract, EOS 69, 51, 1659, 1988

Lagios, E., J. Drakopoulos, R.G. Hipkin and C. Gizeli: Microgravimetry in Greece: applications to earthquake and volcano-eruption prediction, Tectonophysics 152, 197-207, 1988

Roth, F.: Modelling of stress patterns along the western part of the North Anatolian Fault Zone, Tectonophysics 152, 215-226, 1988

Richards, M.A. and R.W. Griffiths: Deflection of plumes by mantle shear flow: experimental results and a simple theory, Geophys. Journal 94, 367-376, 1988

Romaides, A.J., Ch. Jekeli, A.R. Lazarewicz, D.H. Eckhardt and R.W. Sands: A detection of Non-Newtonian Gravity, J.G.R. 94, B2, 1563-1572, 1989

Rubincam, D.P., B.F. Chao and K.H. Schatten: Non-Newtonian gravity and gravity anomalies, EOS 69, 50, 1636, 1988

Schwarzschild, B.: From mine shafts to cliffs - The 'fifth force' remains elusive, Physics Today 41, 7, 1988

Speake, C.C. and T.J. Quinn: Search for a Short-Range, Isospin-Coupling Compo-
    nent of the Fifth Force with Use of a Beam Balance, Physical Review Let-
    ters 61, 12, 1340-1343, 1988
Stubbs, C.W., E.G. Adelberger, B.R. Heckel, W,F. Rogers, H.E. Swanson, R. Wa-
    tanabe, J.H. Gundlach and F.J. Raab: Limits on Composition-Dependent In-
    teractions Using a Laboratory Source: Is There a "Fifth Force" Coupled to
    Isospin? Physical Review Letters 62, 6, 609-616, 1989

Accuracy of GPS in Crustal Deformation Studies: Observation and Adjustment Design

D.B. Grant
Department of Survey & Land Information
Wellington, New Zealand

ABSTRACT:

The technique of covariance analysis has been frequently used to determine the effect of systematic errors on GPS derived coordinates. In this paper, covariance analysis is applied to GPS crustal deformation surveys. The GPS systematic errors are propagated through to the parameters of a subsequent deformation adjustment. Three GPS observation and adjustment schemes with varying accuracy are considered. Deformation adjustments are simulated using terrestrial data in one epoch and GPS data in another epoch. It is demonstrated that the GPS systematic errors in these examples are less important than the terrestrial errors. Deformation adjustments using GPS data in both epochs are also studied. In these cases, systematic errors exceed the formal errors derived from the adjustment. The value of covariance analysis for improved statistical testing of the significance of deformation is demonstrated.

1 INTRODUCTION

The results presented in this paper are based on simulated data. Simulations allow us to test observation and adjustment scenarios, to generalise about the effect of systematic errors and to improve statistical testing of the results of an adjustment. Least squares estimation provides a "maximum likelihood" solution if observation errors are random. With GPS the random component of errors is generally very small and systematic errors play a major role. Thus the formal errors of the parameters in a GPS adjustment are usually very optimistic. These define the precision of the adjustment but are a poor indication of the accuracy.

A technique often used in satellite geodesy is covariance analysis. The adjustment model is extended to include systematic parameters. Given reasonable hypothesized magnitude and behaviour for the systematic errors, we then obtain an indication of the accuracy of the adjusted parameters including the coordinates. While this is of interest, the coordinates are often not the final product of deformation analysis. For example, they may be used in a subsequent adjustment to derive the parameters of rotation, dilatation and strain. Thus it is useful to propagate the systematic errors through to the derived deformation parameters. This allows us to develop optimum observation and adjustment strategies for deformation surveys. It allows us to identify the most important systematic errors for specific applications and allows more reliable testing of the statistical significance of deformation.

2 DEFORMATION

The deformation model used is that of infinitesimal homogeneous strain. This model has proved useful in the past although the assumption of homogeneity can only be considered valid for areas covering a few tens of kilometres in New Zealand (Bibby and Walcott, 1977). A topocentric coordinate system is used with the axes defined by the east, north and vertical (E,N,V) directions at a defined point near the centroid of the network.

The 12 parameters of this model are the 3 translations (t_e , t_n , t_v), 3 rotations (ω_e , ω_n , ω_v) and the 6 elements of the symmetric strain tensor S .

$$S = \begin{pmatrix} e_{ee} & e_{en} & e_{ev} \\ e_{en} & e_{nn} & e_{nv} \\ e_{ev} & e_{nv} & e_{vv} \end{pmatrix} \qquad (1)$$

In practice the vertical strain parameters (e_{ev} , e_{nv} , e_{vv}) are not estimated because they are poorly defined in a geodetic network in which all points lie within 1 - 2 km of a horizontal plane. For further discussion of the estimation of vertical strain parameters, see Grant (1988).

From the horizontal strain parameters the dilatation Δ , components of shear strain γ_1 and γ_2 , and total shear γ can be derived (e.g. Prescott et al, 1978).

$$\text{Dilatation} \qquad \Delta = e_{ee} + e_{nn} \qquad (2)$$

$$\text{Pure Shear} \qquad \gamma_1 = e_{ee} - e_{nn} \qquad (3)$$

$$\text{Engineering Shear} \qquad \gamma_2 = 2e_{en} \qquad (4)$$

$$\text{Total Shear} \qquad \gamma = (\gamma_1^2 + \gamma_2^2)^{1/2} \qquad (5)$$

In the following simulations, variance-covariance (VCV) matrices in (Δ , γ_1 , γ_2) are derived. The variance of γ cannot be determined without knowledge of the magnitudes of γ_1 and γ_2 which are not known in a simulation. However the maximum and minimum variances of γ can be determined by eigenvalue decomposition of the VCV matrix of the subvector (γ_1 , γ_2). Similarly the maximum and minimum variances in rotation about a horizontal axis are given by eigenvalue decomposition of the VCV matrix of the rotations (ω_e , ω_n).

In this paper 4 parameters are used to study the accuracy of deformation adjustments. These are:

R_h Rotation about a horizontal axis. The maximum error in this parameter (determined by eigenvalue decomposition as noted above) indicates the accuracy of the determination of relative uplift or subsidence. Good heights are required in 2 epochs for this parameter to be useful.

R_v Rotation about a vertical axis. The error in this parameter indicates the accuracy of the determination of relative rotation between regions in a heterogeneous strain field. A good definition of orientation is required in both epochs for this parameter to be useful.

Δ   Dilatation. This parameter assists in distinguishing between tension and compression. The estimation of dilatation requires an accurate definition of scale in both epochs to be useful.

γ   Total shear strain. The maximum error in this parameter is determined by eigenvalue decomposition as noted above. This parameter indicates the horizontal distortion of the network. It can be derived from triangulation data where there is no accurate definition of orientation or scale.

3   COVARIANCE ANALYSIS

The equations of covariance analysis may be found in a number of publications including Hatch & Goad (1973) and Bierman (1977). The equations used for this study are derived in Grant (1988). Basically, the vector of adjusted parameters $\mathbf{x}$ is extended to include other parameters $\mathbf{s}$ which are imperfectly known but which are unadjusted. Errors $\delta\mathbf{s}$ in the *a priori* values of these parameters are systematic errors which cause perturbations $\mathbf{p}$ in the estimated corrections $\delta\hat{\mathbf{x}}$ to the adjusted parameters. We distinguish between the computed corrections $\delta\hat{\mathbf{x}}^c$ and the actual corrections $\delta\hat{\mathbf{x}}^a$ that would have resulted if the true values $\mathbf{s}$ of the unadjusted parameters had been used in the adjustment. The perturbations are defined as

$$\mathbf{p} = \delta\hat{\mathbf{x}}^c - \delta\hat{\mathbf{x}}^a \tag{6}$$

Given a model for the behaviour of the systematic errors we can derive the sensitivity matrix $\mathbf{C}$ which allows us to calculate the perturbations resulting from specific systematic errors.

$$\mathbf{p} = \mathbf{C}\,\delta\mathbf{s} \tag{7}$$

Given also, an *a priori* VCV matrix of the systematic errors $\mathbf{Q_s}$, we can derive the VCV matrix of the perturbations $\mathbf{Q_p}$ and thence the actual VCV matrix of the adjusted parameters $\mathbf{Q_{\hat{x}}^a}$ from the computed VCV $\mathbf{Q_{\hat{x}}^c}$.

$$\mathbf{Q_p} = \mathbf{C}\,\mathbf{Q_s}\,\mathbf{C}^T \tag{8}$$

$$\mathbf{Q_{\hat{x}}^a} = \mathbf{Q_{\hat{x}}^c} + \mathbf{Q_p} \tag{9}$$

Where groups of systematic errors are independent we may decompose the perturbation VCV matrix $\mathbf{Q_p}$ and generate a VCV for each group of systematic errors (e.g., orbits, troposphere etc.) that we may wish to consider. This allows us to determine their contribution to the actual or total error.

The main programs used in this study are DASH and COSTRAIN described in Grant (1988). Briefly, DASH is a general purpose covariance analysis program which accepts partial derivatives for a wide variety of parameters and simulates a Kalman filter adjustment in which some parameters are adjusted. Postulated *a priori* errors in the unadjusted parameters are used to generate perturbations and perturbation VCVs. DASH is used in this study to simulate single-session, multi-station adjustments of GPS carrier phase data. The output of several simulated session adjustments from DASH may be combined in COSTRAIN to simulate an adjustment of homogeneous strain parameters. Another program, CONET, can be used to simulate a network adjustment. In COSTRAIN and CONET, the perturbations and perturbation VCVs are again computed giving the effect of GPS systematic errors on the strain parameters or network coordinates.

4   NETWORK OBSERVATIONS AND ADJUSTMENT

The network considered in this paper is part of the New Plymouth - Castle Point Earth Deformation Studies (EDS) network, New Zealand shown in Figure 1.

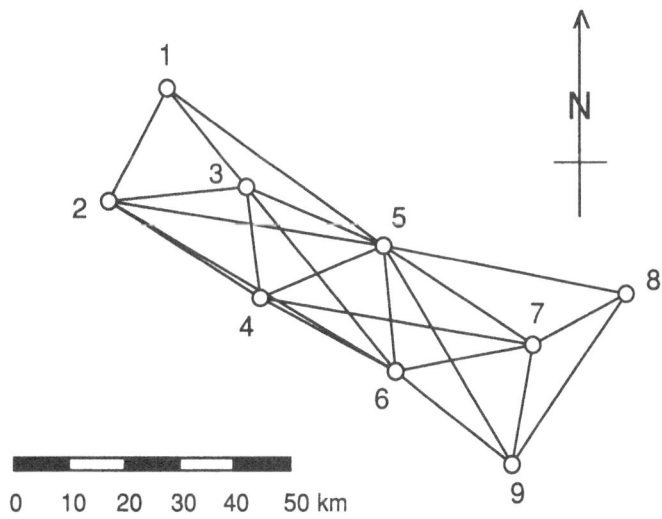

FIGURE 1   Sub-network of the Castle Point - New Plymouth Earth Deformation Studies network, North Island, New Zealand.

Good quality terrestrial data is available for this network and is described in Hannah (1986). The terrestrial data consists of horizontal directions, EDM distances, vertical angles and astronomic observations of latitude, longitude and azimuth. This data has been adjusted using the integrated geodetic program OPERA described in Landau et al (1988). For details of the 3D adjustment, see Grant (1988). The VCV matrix from this adjustment is used in simulating the deformation adjustments.

A simulated GPS survey of this network has also been generated in which 4 receivers occupy the 9 network stations in 6 observing sessions. This is based on an 18 satellite constellation. Details of the GPS observation scheme are given in Grant (1988).

Three different GPS adjustment scenarios are used here:
1. Standard GPS Survey.
   – Broadcast ephemeris used with an assumed accuracy of ±20 metres.
   – Origin station coordinates given by the C/A code point position with an assumed accuracy of ±10 metres in each component.
   – Single frequency observations at night with a peak ionospheric zenith delay error of ±1.5 metres. This assumes a ±3 metre zenith delay with correction to 50% given by the Klobuchar model (Klobuchar, 1986).
   – Tropospheric delay modelled using surface meteorological observations with a residual error of ±2% of the total delay.
2. Dual Frequency GPS Survey
   – As above but with ionospheric error eliminated through the use of dual frequency data.
3. High Accuracy GPS Survey
   – Precise ephemeris from some source with assumed accuracy of ±2 metres.
   – Origin coordinates from Doppler observations with assumed accuracy of ±3 metres in each component.
   – Ionospheric error eliminated through the use of dual frequency data.
   – Tropospheric zenith delay parameter estimated for each station as a 1st order Gauss Markov process with surface meteorological observations providing *a priori* estimates. (See e.g., Lichten and Border, 1987).

Ambiguities are resolved in the simulated adjustment if the total error in the ambiguity is less than 0.5 cycles at the 99% confidence level. This is an iterative process which continues until all ambiguities are resolved or until all remaining ambiguities fail the resolution criterion. In the standard GPS survey 58 of the 90 ambiguities can be resolved using this criterion. In the dual frequency survey 74 ambiguities are resolved and in the high accuracy survey all 90 ambiguities are resolved.

5 DISCUSSION OF RESULTS

5.1 Terrestrial / GPS Deformation Adjustment

The first data sets generated use a terrestrial observation scheme for epoch 1 and simulated GPS schemes for epoch 2. The formal errors in the deformation adjustment are based on the computed VCV matrices of the terrestrial and GPS network adjustments. Perturbations to the deformation parameters are computed for GPS systematic errors. Systematic errors in the terrestrial data are not considered. In this case the formal coordinate errors of the terrestrial network adjustment are 1 to 2 orders of magnitude greater than those of the GPS adjustments. Thus the formal error of the deformation parameters is almost completely dependent on terrestrial data errors.

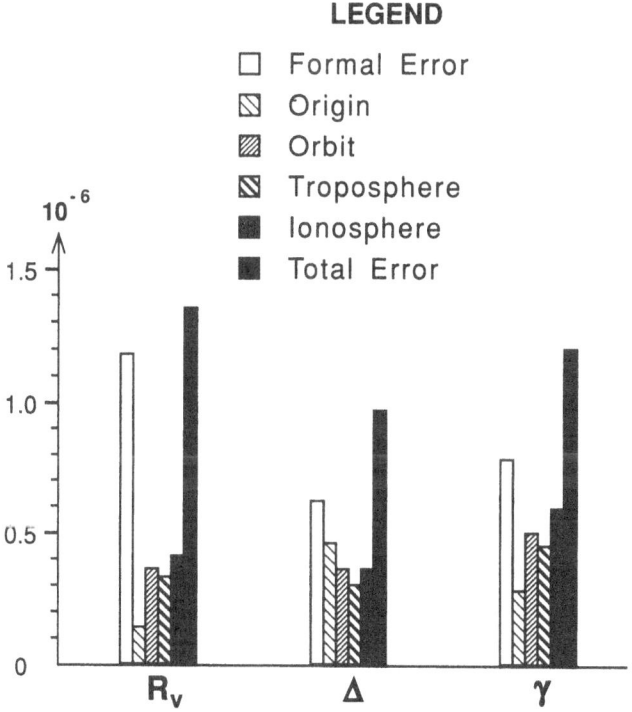

FIGURE 2   Error budget for a deformation adjustment combining data from a terrestrial survey in epoch 1 and a standard GPS survey in epoch 2.

Figure 2 shows the effect of GPS systematic errors on parameters of a deformation adjustment. The height errors in the terrestrial adjustment are such that the estimate of

R_h (rotation about a horizontal axis) is of no value for deformation studies and is not shown. For the 3 parameters R_v, Δ and γ, the formal error dominates the total error budget. As noted above, the formal error is itself mainly dependent on the terrestrial observation errors. Thus even with the standard GPS survey, GPS errors, random or systematic, do not play a major role in the deformation adjustment. Where the terrestrial data is even less accurate (which is the case in New Zealand with the terrestrial surveys of last century) the accuracy of a standard GPS survey will be more than sufficient for deformation analysis.

Figure 3 shows the total error in the deformation parameters for the 3 chosen levels of GPS survey accuracy. It is seen that, due to the relatively large terrestrial survey errors, higher GPS accuracy leads only to limited improvement in the accuracy of the deformation parameters. The greatest improvement is in the estimate of γ. The high accuracy GPS survey gives an estimate of γ 35% more accurate than that resulting from the standard GPS survey.

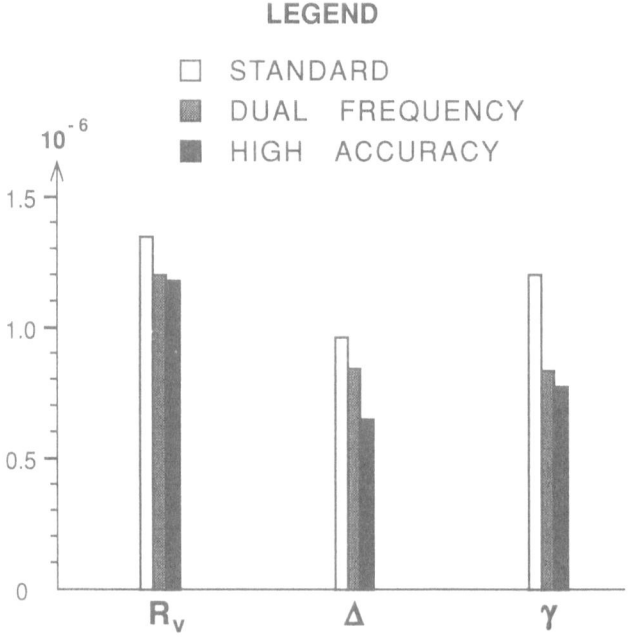

LEGEND

☐ STANDARD
▨ DUAL FREQUENCY
▧ HIGH ACCURACY

FIGURE 3   Total errors for deformation adjustments combining data from a terrestrial survey in epoch 1 and each of 3 different GPS survey adjustments in epoch 2.

Figure 4 shows the normalised GPS perturbations for the 3 levels of GPS survey accuracy. The normalised perturbation is the total perturbation due to GPS systematic errors divided by the formal error. Ideally this should be significantly less than unity to justify the assumption of normally distributed observation errors. The reduction in the significance of perturbations as the accuracy of the GPS survey increases is seen in this figure. For the standard GPS survey the total perturbation in Δ and γ exceeds 1σ. For the high accuracy GPS survey the total perturbation is 0.05σ for R_v, 0.32σ for Δ and 0.13σ for γ. Thus the GPS systematic errors in the high accuracy GPS survey are swamped by terrestrial data errors.

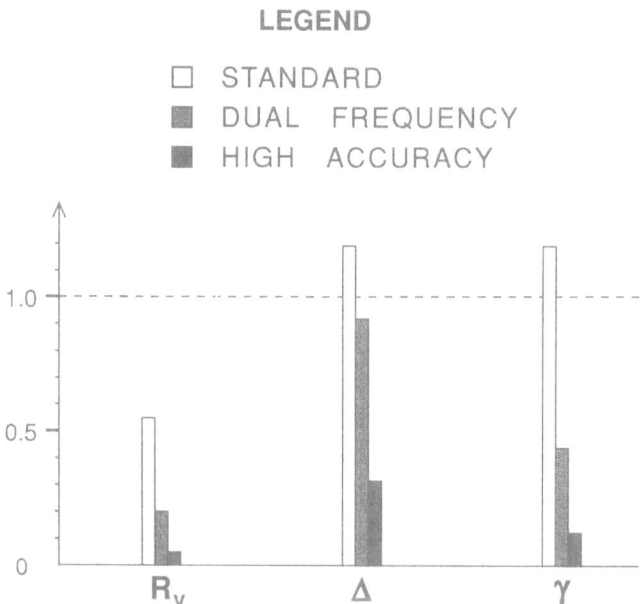

FIGURE 4   Normalised perturbations (total perturbation divided by formal error) for deformation adjustments combining data from a terrestrial survey in epoch 1 and each of 3 different GPS survey adjustments in epoch 2.

5.2 GPS / GPS Deformation Adjustment

Figures 5, 6 and 7 show results equivalent to figures 2, 3 and 4 for the case where GPS observations are available in both epochs. The perturbations shown in these figures are those due to the systematic errors of 1 epoch only. However the formal errors, as above, are based on the VCV matrices of the coordinates in both epochs. Identical GPS observation and adjustment schemes are used for both epochs.

One notable feature of figure 5 is that the formal errors in the GPS / GPS deformation adjustment are very small compared with the total error budget. This is a common feature of GPS adjustments, particularly where atmospheric and orbital parameters are not estimated. The dominant error in R_h is due to the troposphere. It is well known that an error in the differential tropospheric delay between stations leads to a differential height error. The poor definition of height due to geometry and tropospheric error leads to large perturbations in the estimate of R_h.

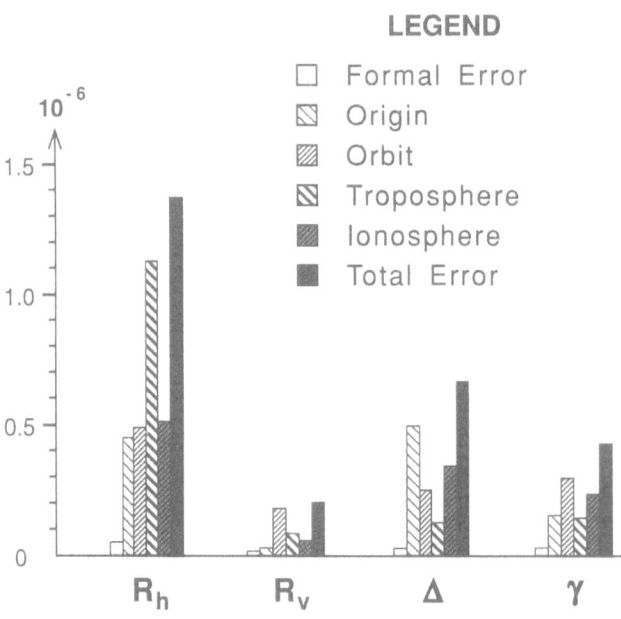

FIGURE 5   Error budget for a deformation adjustment combining data from standard GPS surveys in epochs 1 and 2.

The largest error source in R_v is the orbits. This is expected as the orientation of a GPS network (with unadjusted orbital parameters) is defined by the satellite ephemerides. The main error sources in the estimate of Δ are origin coordinates and the ionosphere. Beutler et al (1988) demonstrate that an error in the height of the origin station causes a scale error. It is also well known (e.g., Beutler et al, 1988) that an ionospheric delay error common to all stations causes a scale error. Thus these results are in accord with expected behaviour. The dominant error sources in the estimate of γ are orbits and ionosphere. As the simulated observation scheme covers 6 sessions, any changes in orientation (orbits) or scale (orbits and ionosphere) between sessions will tend to distort the network and lead to shear strain. Note that the character of errors in a multi-session scheme will therefore be different from those of a single-session scheme. This highlights the advantage of being able to simulate a multi-session adjustment.

In figure 6 the accuracy of the deformation parameters is shown for the 3 GPS adjustment models. The errors of the standard GPS survey are 3 to 7 times those of the high accuracy survey. Unlike the terrestrial / GPS adjustment, systematic errors are major error sources and their elimination or reduction leads to significant improvement in the GPS / GPS deformation adjustment. For the high accuracy GPS survey, the total error in R_h is 0.45 μrad (0.09 arcsec), the error in R_v is 0.03 μrad (0.006 arcsec), the error in Δ is 0.20 μstrain and for γ it is 0.07 μstrain. Note however that this includes the systematic errors of 1 epoch only.

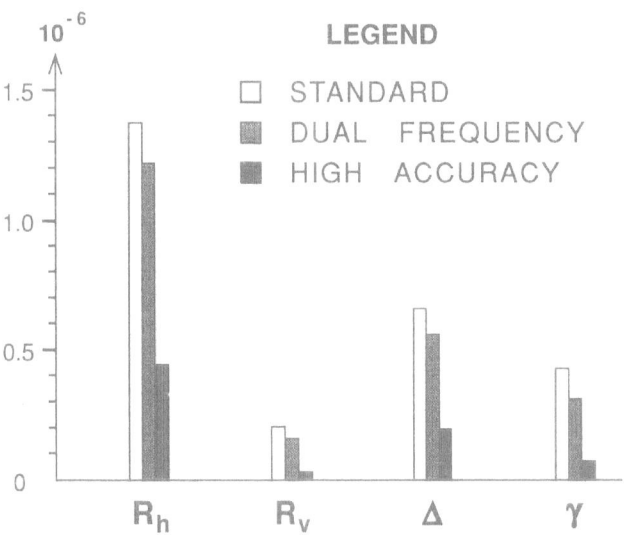

FIGURE 6   Total errors for deformation adjustments where GPS data is available in both epochs. Results for each of 3 different GPS adjustment models are shown.

The effect of tackling only one of the major error sources is given by comparing the standard and dual frequency surveys. The improvement in accuracy for the dual frequency survey ranged from 11% (R_h) to 27% (γ). Whether this improvement justifies the increased cost depends on the requirements of the deformation study. The best results are naturally obtained when all major systematic error sources are addressed as in the high accuracy GPS survey. Note in this case that the ionospheric error assumed was for night time observations at a time of high solar activity. Day-time observations would show greater ionospheric errors and the improvement resulting from the use of dual frequency would be greater.

Figure 7 shows the normalised perturbations for the GPS / GPS deformation adjustment. The total systematic perturbation exceeds 2σ in all cases and thus the assumption of normally distributed errors is not valid. The normalised perturbations of the standard GPS survey are 4 to 12 times those of the high accuracy GPS survey. This is due to two effects:

1. The systematic errors have been reduced in the high accuracy GPS survey.
2. The formal errors are greater in the high accuracy GPS survey due to the inclusion of extra parameters in the adjustment (the zenith tropospheric delays).

Thus the high accuracy survey is more accurate but less precise than the standard GPS survey. This narrows the gap between accuracy and precision.

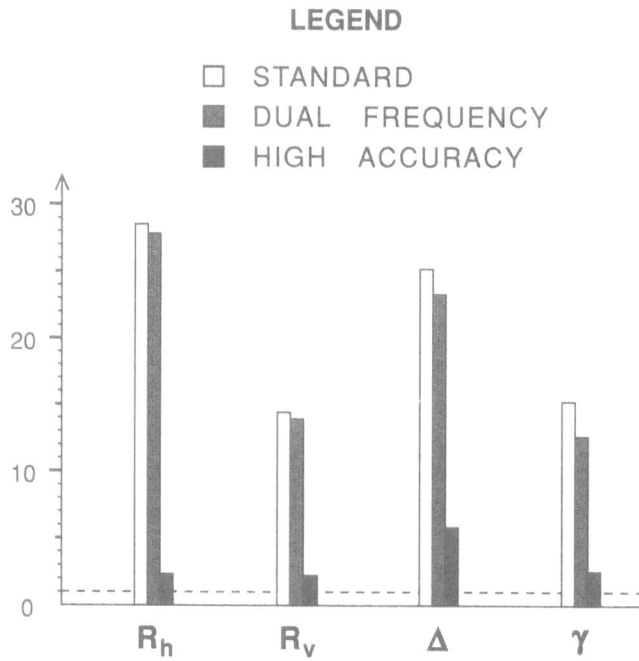

FIGURE 7   Normalised perturbations (total perturbation divided by formal error) where GPS data is available in both epochs. Results for each of 3 different GPS adjustment models are shown.

The size of the normalised perturbations demonstrates the value of covariance analysis in testing the statistical significance of estimated deformation parameters. Significance tests based on the formal error may give misleading results. The use of a single variance scaling factor (*a posteriori* variance factor or standard error of unit weight) can improve this testing provided it accurately reflects the level of systematic error. However this can also give misleading results as some parameters (e.g., R_h and Δ) have higher normalised perturbations than others (e.g., R_v and γ).

6 CONCLUSIONS

The high accuracies possible with GPS have encouraged research in the use of GPS for earth deformation studies. This has concentrated on the goal of detecting deformation in years or months that would have taken decades using terrestrial techniques. However we should not ignore the wealth of terrestrial data collected over the last 100 years or more in active deformation zones. GPS data can be combined with historical terrestrial data to increase our knowledge of deformation. It is demonstrated in this paper that the GPS data need not be of the highest quality to be of value. GPS surveys conducted for other purposes using single frequency receivers and standard processing models can be usefully combined with terrestrial data. The principal difficulty in combining terrestrial and GPS data for deformation studies is the relatively high errors, random and systematic, in the terrestrial data.

Looking to the future, deformation can also be studied using repeated GPS surveys. Depending on the level of accuracy required, GPS systematic errors may be of concern. The most difficult task in deformation analysis is not estimating the parameters but determining their significance. The formal errors of GPS adjustments are typically too small to represent the accuracy of the data. Covariance analysis as demonstrated in this paper can be used to gain a better understanding of the accuracy of the deformation parameters. This is dependent on the assumed magnitude and behaviour of the systematic errors. While this places a constraint on the value of covariance analysis, the assumptions made will generally be better than the common assumption that observation errors are completely random.

Another role of covariance analysis is in the design of observation schemes and adjustment models. Deformation surveys are expensive and it is important to optimise the data collected within budget restraints. For example, a decision on whether to use dual frequency receivers can be based on the expected improvement in accuracy. It may be that the expense of dual frequency observations cannot be justified unless other major error sources are also dealt with. Covariance analysis provides the information needed for such decisions.

If the final product of the deformation analysis is coordinate displacements (e.g., deformation surveys of structures such as dams) perturbations in the adjusted coordinates are of interest. However, if the coordinates are to be used in a subsequent deformation adjustment to derive other parameters such as rotation or strain, the extended covariance analysis demonstrated in this paper is more appropriate.

7 REFERENCES

Beutler G, Bauersima I, Gurtner W, Rothacher M, Schildknecht T, Geiger A (1988) Atmospheric refraction and other important biases in GPS carrier phase observations., In Brunner FK (ed), Atmospheric effects on geodetic space measurements, Monograph 12, School of Surveying, University of N.S.W., Sydney, pp15-43.

Bibby HW, Walcott RI (1977) Earth deformation and triangulation in New Zealand., New Zealand Surveyor 252: pp741-762.

Bierman GJ (1977) Factorization methods for discrete sequential estimation., Academic Press, Orlando.

Grant DB (1988) Combination of terrestrial and GPS data for earth deformation studies in New Zealand., PhD thesis, School of Surveying, University of N.S.W., Sydney.

Hannah J (1986) Accuracy estimates of, and the rejection criteria used on the observed quantities from the New Zealand EDS networks., In Reilly WI, Harford BE (eds) Recent crustal movements of the Pacific region, Bull. 24, Royal Society of New Zealand, pp463-471.

Hatch W, Goad C (1973) Mathematical description of the ORAN error analysis program., Wolf Research & Development Corp., Riverdale, Maryland.

Klobuchar JA, (1986) Design and characteristics of the GPS ionospheric time delay algorithm for single frequency users., In Proceedings of PLANS 86, Position Location and Navigation Symposium, Las Vegas, November 1986, pp280-286

Landau H, Hehl K, Eissfeller B, Hein GW, Reilly WI (1987) Operational geodesy software packages., Institute of Astronomical and Physical Geodesy, University of Federal Armed Forces, Munich.

Lichten SM, Border JS (1987) Strategies for high precision GPS orbit determination., Journal of Geophysical Research, 92: pp12751-12762.

Prescott WH, Savage JC, Kinoshita WT (1979) Strain accumulation rates in the western United States between 1970 and 1978., Journal of Geophysical Research, 84: B10, pp5423-5435

Monitoring Crustal Motion in Papua New Guinea Using the Global Positioning System

A. Stolz
School of Surveying
University of New South Wales
Australia

ABSTRACT:

Papua New Guinea is a region of intense and frequent earthquake activity. Four earthquakes of magnitude 8.0 or greater have occurred here in the last century. There are about 100 major volcanoes of which 14 are classified as active and 24 are classified as dormant. Bouguer gravity anomalies range from about -180 mGal to 200 mGal and the minimum free-air gravity anomalies reach about -300 mGal. The region occupies a unique position on the global satellite geoid - at the crest of a bulge which is higher than other parts of this geoid. Papua New Guinea is also a region of large predicted plate tectonic motions. The region includes two, and possibly as many as four, minor plates sandwiched between the major Indo-Australian and Pacific plates.

It is proposed to use Global Positioning System (GPS) receivers to monitor crustal motion in Papua New Guinea. Because of the location of various islands on both sides of the plate boundaries, and within the broad inter-arc area, it is possible to use GPS to establish baselines that straddle many of the major tectonic elements, and so by repeated observations of these baselines, to directly observe the kinematics of plate convergence, intra-arc strain and back-arc spreading. Because the rates of plate convergence and back-arc spreading in this region are among the highest found in the world, and baseline length is generally under 500 km, it should be possible to attain an unusually high ratio of tectonic signal-to-measurement noise.

This paper summarises the principal tectonic elements of the Papua New Guinea region, briefly reviews the results of previous satellite surveys established to monitor crustal deformation in the region, elucidates the main scientific objectives of a comprehensive program to monitor crustal deformation in Papua New Guinea using GPS and describes a modest campaign of first-epoch GPS measurements to be observed at the earliest opportunity, possibly in 1990. This initial GPS survey, viewed as a proof-of-concept campaign, would provide a high-accuracy framework for future crustal motion monitoring activity in Papua New Guinea. By reoccupying stations observed in 1981 with Doppler receivers, a preliminary determination of tectonic motion is possible after one GPS campaign.

1. INTRODUCTION

One of the major developments in the Earth Sciences in recent years has been the formulation of the plate tectonic hypothesis. This hypothesis has had a profound influence on all geological thinking over the past two decades and has placed order in a seemingly maze of disparate geological, geophysical and geochemical observations. It is providing a framework for understanding the occurrence of earthquakes and volcanic eruptions and their associated activity, as well as a strategy for mineral and hydrocarbon exploration. The plate tectonics models are based on a variety of geophysical and geological observations which have limited spatial or temporal resolution and many aspects of the models remain to be tested. To what extent do the tectonic plates behave as rigid entities? Are the plates moving uniformly? How are the motions between the major plates taken up at the plate boundaries? Geodetic observations are now able to answer some of these questions. By measuring crustal displacements it is possible to deduce strain and infer the stress fields that produce the deformation. It becomes possible to measure "instantaneous" motions over intervals of a few years or less and it becomes possible to define zones over which the deformation occurs. One of the major areas of progress in satellite geodesy that has made this possible is the Global Positioning System (GPS). The Australian region is particularly suited for these geodetic studies. Relative plate motions are high, of the order of 10 cm/a, and plate boundaries are accessible for these studies. The region chosen for these studies is that of Papua New Guinea where preliminary work has already been carried out and where there has been considerable Australian involvement in geological and tectonic exploration.

We propose to use GPS receivers to monitor crustal motion in Papua New Guinea. Here the major Indo-Australian and Pacific plates converge. The region includes two, and possibly as many as four, minor plates sandwiched between the major plates. There are at least six plate boundaries. Most of them are zones of convergence, characterised by different components of strike-slip motion; several are ridge-transform zones where new seafloor is being created (Johnson, 1979). Rates of plate tectonic motion are high, estimated to be about 13 cm/a across the New Britain Trench (Fig. 1). The precise nature of these movements, possibly intermediate between motions along discrete plate boundaries and deformations uniformly distributed over a broad area, remains a major unsolved problem despite several decades of intensive research into the tectonics of this region (eg. Johnson and Molnar, 1972; Curtis, 1973; Krause, 1973; Taylor, 1975; Hamilton, 1979; Johnson, 1979; Taylor, 1979; Honza et al., 1987; Taylor et al., 1986; Marlow et al., 1988). The Bismarck Sea is floored by an actively spreading back-arc basin (Connelly, 1976; Taylor, 1979). The detailed nature of back-arc spreading processes and their driving forces have not yet been identified.

GPS has recently emerged as a very effective and relatively inexpensive tool for monitoring crustal motions at the regional distance scales. Currently, GPS baseline measurements can be made with an accuracy of about 0.1 ppm over distances of up to about 1 000 km (Bertiger and Lichten, 1987; Beutler et al., 1987; Tralli et al., 1988; Dong and Bock, 1989). Because of the location of various islands on both sides of the plate boundaries, and within the broad inter-arc area, it is possible to use GPS to establish baselines that straddle many of the major tectonic elements, and so by repeated observations of these baselines, to directly observe the kinematics of plate convergence, intra-arc strain and back-arc spreading. Because the rates of plate convergence and back-arc spreading in this region are among the highest found in the world, and baseline length is generally under 500 km, it should be possible to attain an unusually high ratio of tectonic signal-to-measurement noise.

A minimum of two GPS measurement campaigns would normally be required to determine crustal motion. However, a regional survey using the satellite Doppler technique, to establish a framework of widely spaced but accurately positioned stations that could be used in future crustal movement surveys, was performed in 1981 (Angus-Leppan et al., 1983). Results recently obtained by Morgan (P.J. Morgan, 1989, personal communication) indicate baseline measurements which are accurate to about 30 cm. In some areas accumulated tectonic motions should exceed this value, if we assume that the motion is uniform. GPS is at least an order of magnitude more accurate than the Doppler technique. Thus, by reoccupying stations of the Doppler network a preliminary determination of tectonic motion is possible after only one GPS campaign. The GPS survey would provide a high-accuracy framework for future crustal motion monitoring activity in Papua New Guinea.

2. GENERAL TECTONICS OF THE PAPUA NEW GUINEA REGION

Figure 1 shows the generalized tectonics of the Papua New Guinea region. The tectonic framework of the region is dominated by (1) collision of the Australian continent with an island-arc system now partly incorporated in New Guinea, (2) collision of the Ontong Java Plateau with the Solomon island arc, and (3) formation of the active Woodlark and Manus marginal basins (Johnson, 1987). The Bismarck Sea is considered to be an actively spreading back-arc basin (Connelly, 1976; Taylor, 1979). The northern part of the Bismarck Sea possibly belongs to the Pacific plate, or is a poorly defined single platelet bordered by the Manus Trench to the north (Fig. 1). The Manus Trench is a subduction zone of the Pacific plate and is thought to be active at a very slow subduction rate or almost inactive (Weissel and Anderson, 1978). In the Manus Basin, located in the eastern Bismarck Sea, magnetic anomaly data indicate that crustal opening has been progressing along an asymmetric spreading zone for about 3.5 Ma at a total opening rate of about 13 cm/a (Taylor, 1979). A new type of plate boundary called an "extensional transform zone" by

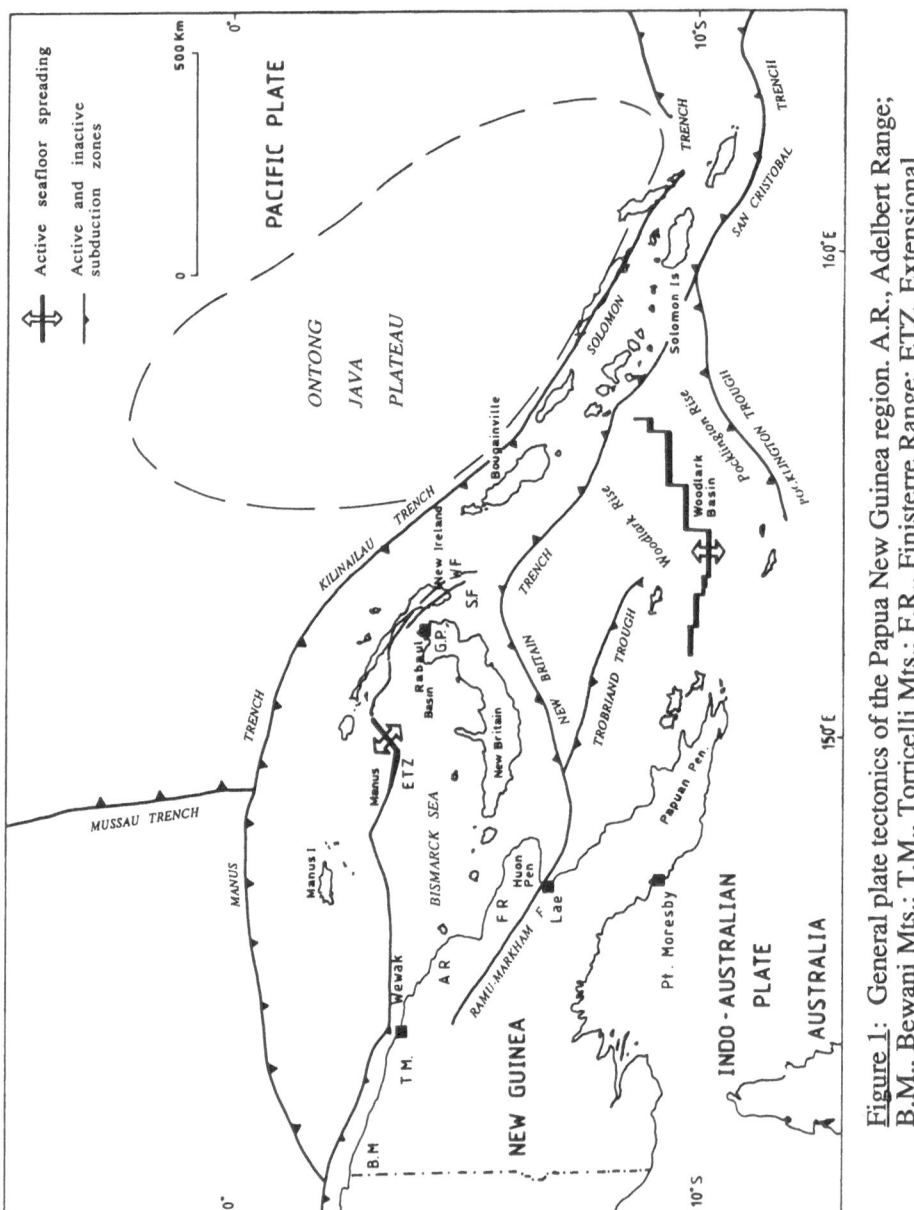

Figure 1: General plate tectonics of the Papua New Guinea region. A.R., Adelbert Range; B.M., Bewani Mts.; T.M., Torricelli Mts.; F.R., Finisterre Range; ETZ, Extensional Transform Zone; G.P., Gazelle Pen.; S.F., Sapom Fault; W.F., Weitin Fault. Triangles on trench segments indicate direction of subduction.

Taylor et al. (1986) has been identified in the Manus Basin. The North Solomon Trench stopped subduction in the Miocene or later (Honza et al., 1987) and the South Solomon Trench and the New Britain Trench started subduction in the Pliocene. The Woodlark Basin consists of a series of north-south transform faults connected by east-west spreading segments. The northeastern edge of the Woodlark Basin is undergoing active subduction along the Bougainville Trench with its active seafloor-spreading axis underthrusting the Pacific plate to the northeast (Taylor, 1987). The Trobriand Trough represents an active subduction zone (Hamilton, 1979; Honza et al., 1987; Lock et al., 1987). Northwards subduction is taking place beneath New Britain to the north, and the region of seafloor beneath the two trench systems is being arched upwards as the trenches converge southeastwards; however, the Trobriand Trough itself appears to be aseismic (Johnson, 1987). Ridge subduction occurs in the late Cainozoic volcanic islands of the New Georgia group (Taylor, 1987). Deformational structures such as upfaulted and rotated blocks of trench-fill sediments which form most of the inner wall of the Kilinailu Trench are evidence for suduction zone origin of the trench and for continuing compression between the Ontong Java Plateau and the island arc (Johnson, 1987). The Finisterre-Huon block, comprising the Finisterre and Saruwaged ranges and Huon Peninsula along the northern margin of Papua New Guinea is considered to have collided with the New Guinea mainland from the north. The present boundary of the collision is thought to be the Ramu-Markham Fault. The block is under compression as suggested by deformation and seismicity (D' Addario et al., 1976; Tingey and Grainger, 1976; Jaques and Robinson, 1976).

There have been numerous assessments of the nature of plate tectonic motions in the Papua New Guinea region (eg. Johnson and Molnar, 1972; Curtis, 1973; Krause, 1973; Taylor, 1975; Hamilton, 1979; Johnson, 1979; Taylor, 1979). A general scheme of plate configurations and relative motions has emerged from these studies, but there are some striking differences in interpretation, caused mainly by the complexity of the plate boundary configurations. Johnson (1979) summarises four of the existing plate tectonic models for the region. A fifth version is given by Hamilton(1979). At least two minor plates are shown between the larger Indo-Australian and Pacific plates in all models. The minor plates are called the North Bismarck, South Bismarck and Solomon Sea plates. Calculated velocity vectors for the plate boundaries are given in Table 1. Relative motions are high of the order of 10 cm/a. Relative movement between the South Bismarck and Solomon Sea plates is nearly orthogonal at very high rates of convergence in the east, decreasing westward. Calculated vectors are 6-12 cm/a. Movement between the Solomon Sea and Pacific plates also is nearly orthogonal at a very high rate of convergence, estimated to be in the range of 7-11 cm/a. Spreading at a rate of 10-13 cm/a between the North Bismarck and South Bismarck plates is taking place in the Manus Basin (Taylor et al., 1986). In the Woodlark Basin, spreading between the Indo-Australian and Solomon Sea plates ranges from 7 cm/a in the eastern part of the basin (Cooper and Taylor, 1987) to 14 cm/a in the western part of the basin (Taylor, 1987).

The boundaries between the Solomon Sea and South Bismarck plates and the Solomon Sea and Indo-Australian plates are zones of particular interest to us and we describe them in some detail below.

3. SOLOMON SEA/SOUTH BISMARCK PLATE BOUNDARY

The Solomon Sea plate is bounded on the north by the New Britain Trench, on the west by a zone of continental rifting and diffuse seismicity in the Papuan Peninsula, and on the south by the Woodlark Basin spreading system (Weissel et al., 1982). The Solomon Sea plate is believed to be converging on and being subducted under the South Bismarck plate along the New Britain Trench. The estimated rate of convergence is high ranging from about 9-13 cm/a (Table 1). Krause(1973) suggested a doubling of the convergence rate northeastwards along the plate boundary, but this effect is much less marked in the results listed by Taylor(1975). Curtis(1973) also proposes greater convergence rates northeastwards, noting an apparent eastward decrease in the dip of the New Britain Benioff zone. This northeastward increase corresponds to progressively deeper parts of the New Britain Trench and to the greater frequency of earthquakes in the northeast. In this region earthquakes occur to depths of 550 km and more.

Table 1

Plate boundary	Hamilton [1979] rate cm/a	Hamilton [1979] azimuth deg	Johnson & Molnar [1972] rate cm/a	Johnson & Molnar [1972] azimuth deg	Curtis [1973] rate cm/a	Curtis [1973] azimuth deg	Krause [1973] rate cm/a	Krause [1973] azimuth deg	Taylor [1975] rate cm/a	Taylor [1975] azimuth deg
I-P	11.0-13.0	70-80	10.0	75	9.5-10.7	80-72	12.3-13.5	68-64	10.7-11.8	73-68
N-P	0.0-3.0	36	1.0	114						
B-N	12.0	97-120	7.5	90						
B-P	14.0	~150	8.4	93	9.4	80-112	13.5	64-112	7.6-8.6	110-131
I-B	0.0-6.0	10	3.3	23	2.6-5.2	18-7.5	0.0-6.2	31	8.0-9.3	29
S-B	6.0-10.0	353-357	9.2	343	12.0	352	6.2 / 12.4	327-5 / 327	11.7 / 12.5	11 / 8
S-P	10.0	56	10.1	34	10.9	40	6.7-7.8	47-48	10.7	51
S-I	4.0	326	7.0	325	7.1	341	2.8-6.5 / 5.6-6.6	287-268 / 276-270	4.1 / 4.1	325 / 332
W-I	0.0-3.0	12								

I = India-Australia plate, P = Pacific plate, N = North Bismark plate, B = South Bismark plate, S = Solomon plate, W = Woodlark plate. Results from Johnson and Molnar [1972], Curtis [1973], Krause [1973], and Taylor [1975] are taken from Table 1 of Johnson [1979]. Variations in plate rates and directions given are not necessarily uncertainties in the motions, but in many cases, are variations in rate and direction at different locations along the plate boundaries.

4. SOLOMON SEA/INDO-AUSTRALIAN PLATE BOUNDARY

The northern and eastern margins of the Solomon Sea plate are well defined by the New Britain Trench, but the southwestern and southern margins, making up the Solomon Sea/Indo-Australian plate boundary, are zones of considerable tectonic complexity, the structure of which is still poorly determined. The earthquake zone of the Solomon Sea/Indo-Australian plate boundary is diffuse and discontinuous (Johnson, 1979). A minor zone of seismicity marking the eastern part of the Solomon Sea/Indo-Australian plate boundary passes along the southern flank of the Woodlark Rise at about 154°E, where it appears to be displaced southeastwards to the axis of the Woodlark Basin, and eastwards intersects the main earthquake zone southeast of Bougainville. There are also epicentres on the northeastern part of the Woodlark Rise and an extensional origin of the Woodlark Basin seems likely (Johnson, 1979). The seafloor spreading characteristics of the Woodlark Basin consist of a series of north-south transform faults connected by east-west spreading segments. The eastern Woodlark Basin is bounded on the south by the Pocklington Rise and Trough, on the northeast by the Woodlark Rise, and on the west by a meridional transform at about 152°E. The northeastern edge of the Woodlark Basin is an active subduction zone along which the youthful basin and its active seafloor spreading axis are underthrusting the Pacific plate to the northeast. The eastern Woodlark Basin is opening at a rate of 7 cm/a and at the same time is being subducted beneath the Solomon Islands, the spreading system forming a triple junction with the New Britain and San Cristobal trenches (Cooper and Taylor, 1987). Southeast of the triple junction the convergence rate is about 11 cm/a along an azimuth of N73°E (Molnar et al., 1975) and northwest of the junction it is about 14 cm/a along N45°E (Taylor, 1987). The controversial interpretation by Hamilton, that the Trobriand Trough represents an active subduction zone, is confirmed by Honza et al. (1987) and Lock et al. (1986). Northwards subduction is taking place beneath New Britain to the north, and the region of the seafloor beneath the two trench systems is being arched upwards as the trenches converge southeastwards. However, the Trobriand Trough is aseismic (Johnson, 1987).

5. MONITORING CRUSTAL PLATE MOTION IN PAPUA NEW GUINEA

Scientific Objectives

The main scientific objectives of a crustal motion monitoring program in Papua New Guinea would be:

(1) Measure the convergence rates across the New Britain Trench. Detection of plate convergence across the New Britain Trench should be possible from two GPS campaigns separated by about one year (but see below). After several years of

monitoring it should be possible to determine convergence rates to within a small uncertainty.

(2) Monitor back-arc spreading in the Manus and Woodlark Basins. This can be achieved directly by repeat measurements of baselines that straddle these areas, and indirectly by determining convergence rates across the arc-trench systems, since these convergence rates are affected by spreading in the inter-arc areas as well as the overall relative motions of the Indo-Australian and Pacific plates.

(3) Monitor the motions on the Pacific plate relative to stations on the Indo-Australian plate. Motions on the Pacific plate relative to stations on the Indo-Australian plate can be compared to predictions of plate motion models such as those of Johnson and Molnar(1972), Curtis(1973), Krause(1973), Taylor(1975, 1979) and Hamilton(1979), and with the results obtained from the NASA Crustal Dynamics Project, in which much longer baselines have been observed using satellite laser ranging and VLBI.

(4) Monitor tectonic strain and uplift in the collision suture of the island of New Guinea where the South Bismarck and Indo-Australian plates converge.

(5) Assess the degree of uniformity associated with plate interactions in the region. Estimates of relative plate motions from plate models such as RM-2, from geological evidence, and from satellite laser ranging and VLBI measurements between continents, assume or suggest that the plates move at a uniform rate relative to one another. The presence of earthquakes, on the other hand, suggests that at some scales the motion is irregular. It may be that the motion of very large plates, such as the Pacific and Indo-Australian plates, is uniform but that the small plates caught between them move irregularly. If this were demonstrated to be the case here, it would have major implications for earthquake prediction efforts and the understanding of the driving forces behind plate motions.

(6) 'Capture' a major earthquake within the geodetic network. The eastern New Britain, southern New Ireland and northern Bougainville regions are among the most seismically active on earth. A high density of stations established in this region would ensure that several of these stations should be close to the rupture zone of the next great earthquake. Measuring the coseismic deformation may allow forecasting of the next event in the region if one assumes that the next earthquake will occur when the strain in the region recovers the same value it had before the measured event.

(7) Determine the extent to which horizontal and vertical motions in different parts of the region are coupled. For example, we expect that horizontal convergence across the New Britain Trench will be accompanied by both vertical motion and the build-up of intra-arc strain across the Baining and Sapom faults in New Ireland. Direct

observations of relationships of this kind can improve our understanding of the earthquake cycle and long-term evolution of the Papua New Guinea region.

(8)   Place additional constraints on global tectonics models. In their RM-2 plate motion model, Minster and Jordan(1978) did not include any data from north of 25°N to determine the location of the Indo-Australia/Pacific plate pole of rotation. The location of the pole is based solely on 14 slip vector azimuths, and a requirement of closure around the Indo-Australian, Pacific and Antarctic plate triple junction. The slip vector azimuths and the zero closure requirement are not mutually compatible, which results in a significant component of compression across the Macquarie Ridge system. Falconer (1973) has hypothesised this to be a purely strike-slip fault system. Minster and Jordan suggest that this incompatibility can be resolved if there is deformation within the Indo-Australian plate. A determination of the present-day plate motion rates using GPS in the Papua New Guinea region could significantly strengthen the estimate of the location and relative rotation rate of the Indo-Australian/Pacific plate pole of rotation.

(9)   Relate the reference levels of tide gauges at Port Moresby and other major ports into a global coordinate system in order to enhance both tectonic deformation monitoring ability in the area and efforts at detecting global sealevel change. Tide gauges can provide a much shorter sampling interval for measuring crustal deformation than annual or biannual GPS measurements and can thus alert us to the presence of shorter term variations in the deformation. Eustatic sealevel is believed to be rising at a rate of about 1 mm/a over the last century (Gornitz et al., 1982). Climatic models indicate that this rate may be accelerating due to recent warming of the atmosphere (Gornitz et al., 1982; Hansen et al., 1981). However, the vertical component of a GPS baseline is the least well determined, and typical accuracies are several centimetres over typical baseline lengths in Papua New Guinea. Thus several decades may have to pass before GPS measurements will be able to contribute significantly to the measurement of global sealevel change.

The above are long-term objectives. Although it should be possible, for example, to detect the rapid plate convergence across the New Britain Trench or the spreading motion in the Manus Basin after just one year of monitoring, the longer the baselines are observed the better one can determine the average plate motion and deformation rates and the extent these vary with time. Achieving a better understanding of the many processes of interest, such as the mechanism of convergence and uplift in the Huon Peninsula region (Chappell, 1974), would require repeated baseline observations over a decade, and perhaps longer.

Previous Crustal Motion Surveys

Markham Valley and St. Georges Channel Terrestrial Surveys

A trilateration network across the Markham Valley, near Lae was observed initially at a time when it was considered that the fault running down the valley formed the boundary between the Indo-Australian and South Bismarck plates (Cook and Murphy, 1974). The network has been remeasured once with no significant motion indicated (Sloane and Steed, 1976a). Possibly these measurements show that the boundary between the Indo-Australian and South Bismarck plates comprises a broad deformation zone, rather than a single fault, and that this deformation zone was not straddled by the survey. Later, a survey network was measured across the St. Georges Channel between New Britain and New Ireland, near Rabaul (Sloane and Steed, 1976b). The eastern boundary of the South Bismarck plate was believed to lie somewhere along this channel. This network was not remeasured to the specifications required for crustal motion surveys. However, a survey party visited the Rabaul area two years prior to the actual crustal motion survey to determine whether some of the longer lines could be measured by geodimeter under tropical conditions. Six test lines were established and measured. Comparisons made between four of the distances observed during these tests with those of the actual survey indicate that no significant movement occurred in the intervening two years. The eastern boundary of the South Bismarck plate is now thought to lie further to the east of the extent of the survey(Mori, 1988). Conventional terrestrial methods were employed for these surveys.

Satellite Doppler Crustal Plate Survey

A regional survey using the multi-station, short-arc satellite Doppler technique to establish a framework of widely spaced but accurately positioned stations that could be used in future crustal movement surveys was performed in 1981. A 15-station network was occupied using five Doppler satellite receivers over a period of about 30 days (Angus-Leppan et al., 1983). Site selection was to a certain extent dictated by logistic considerations. In particular, the number of sites had to be kept below 20 and their location had to be near airfields for easy access. Four stations are located on the South Bismarck plate and two are on the Solomon Sea plate, with a third, close to the plate boundary. There are three stations to the west and south of the Solomon Sea plate, on the Indo-Australian plate. Two stations are located on the North Bismarck plate, and a third may be close to the boundary of this plate. Finally, there are two stations on the Pacific plate, to the east. Atomic frequency standards were employed to increase the accuracy of the results. A multiple reoccupation scheme for three central stations was, moreover, devised to provide for redundancy and ties between the various subnetworks, and a fixed centrally located base station was established. Results recently obtained by Morgan (P.J. Morgan, 1989, personal communication), using the translocation technique to reduce orbital errors and improved models for the large ionospheric disturbances which occurred during the survey, indicate baseline measurements which are accurate to about 30 cm. This network has not been reobserved todate.

Monitoring Crustal Motion in Papua New Guinea with GPS

A comprehensive measurement program aimed at addressing the abovementioned scientific problems would be rather expensive and therefore difficult to justify in these times of economic restraint. However, since plate tectonic motion in the region is large and since a network which could serve as a reference for measurements of relative plate motions is essentially already in place through the 1981 Doppler survey, a modest yet significant start could be made by reobserving all or selected parts of this network with GPS.

We initially propose to determine plate tectonic motion between the Solomon Sea and South Bismarck plates and the Solomon Sea and Indo-Australian plates. This can be achieved by completing one measurement campaign. Normally, a minimum of two measurement campaigns would be required to determine crustal motion. However, since the accumulated tectonic signals across these plate boundaries are expected to exceed the errors in the 1981 Doppler data, we will be able to determine crustal motion after only one GPS campaign. The network which we propose to occupy as soon as possible is shown in Figure 1. The stations of the Doppler survey form the basis of the proposed network. We see this as a proof-of-concept exercise which would also help us in the design of any future expansion of the crustal motion network and in establishing our long-term operational procedure.

In addition, we aim to assess system performance. One of the important technical issues is to determine the minimal configuration of fiducial stations necessary to provide sufficiently accurate orbit determinations. A second important technical issue relates to what impact meteorological data, which provides a deterministic basis for the wet troposphere correction, has on overall system performance. At some sites tropospheric water vapour distribution may be more uniform in space and time than at other sites. Thus, the utility of meteorological observations, particularly water vapour radiometer data (see below), may vary from site to site.

Fiducial Network

In order to measure GPS baselines with an accuracy of about 0.1 ppm it is necessary to determine the satellite orbits to about 1 metre (Stolz et al., 1984). This exceeds normal user requirements by an order of magnitude and we would need to rely on our own facilities to produce orbits of this accuracy. Several methods of independent precision orbit determination exist. In the fiducial network concept, a suite of stations whose locations are accurately known from VLBI or satellite laser ranging define a self-consistent coordinate frame to which all GPS orbit and baseline stations are referred (Davidson et al., 1987). Numerous variations of this concept can be found in the literature. In some cases, a simultaneous solution of orbits, non-fiducial station coordinates and other parameters are obtained (Lichten and Border, 1987). A second approach involves first determining a precision ephemeris with the fiducial stations held fixed and then using this ephemeris to solve for the non-fiducial station positions (Abbot et al., 1986; Bock et al., 1986). A third approach, which does not use fiducial stations, is the free network technique. Here the coordinates of only one site need be held fixed during the orbit and baseline estimation scheme (Beutler et al., 1987). However, common practice in the free network approach is to constrain more than one station and in this way maintain the reference coordinate system. All these methods produce high-precision results. Precision atomic clocks are required in the fiducial network method. The free network method, on the other hand, demands that the satellites are mutually visible from the tracking stations.

The Cooperative International GPS Network (CIGNET), comprising tracking stations in North America (including Hawaii), Europe and Japan, continuously tracks the GPS satellites, providing a fiducial data set that is used for global orbit computation. This fiducial network is suitable for the computation of orbits which are accurate to about 1 ppm globally. Higher accuracy orbits are obtainable from more densely spaced regional tracking networks. The fiducial network proposed for the MIT and NCSU campaigns is described by Bevis and Bock (1989). This fiducial network in addition to CIGNET would be more than adequate for our purposes and we would plan to cooperate with the NCSU and MIT teams and CIGNET so that both the high-density free network and the fiducial network methods of generating high-precision orbits could be tested in Papua New Guinea.

Crustal Motion Network

Figure 2 shows the proposed crustal motion network. This network was designed taking into account our immediate scientific objectives, the availability of GPS receivers, the need to establish a certain redundancy of measurement, logistics and cost. Monuments established during the 1981 Doppler crustal plate survey would be occupied. In this way we would be able to compare the GPS measurements with prior geodetic observations as well as reduce costs.

Several baselines cross the New Britain Trench. The line between Woodlark Is. and Jacquinot Bay on the southeast coast of New Britain is the longest of these lines (about 600 km). If we could attain a relative accuracy of 0.1 ppm, a typical baseline uncertainty is 6 cm. This line is expected to shorten by about 9-13 cm/a. The movement between the South

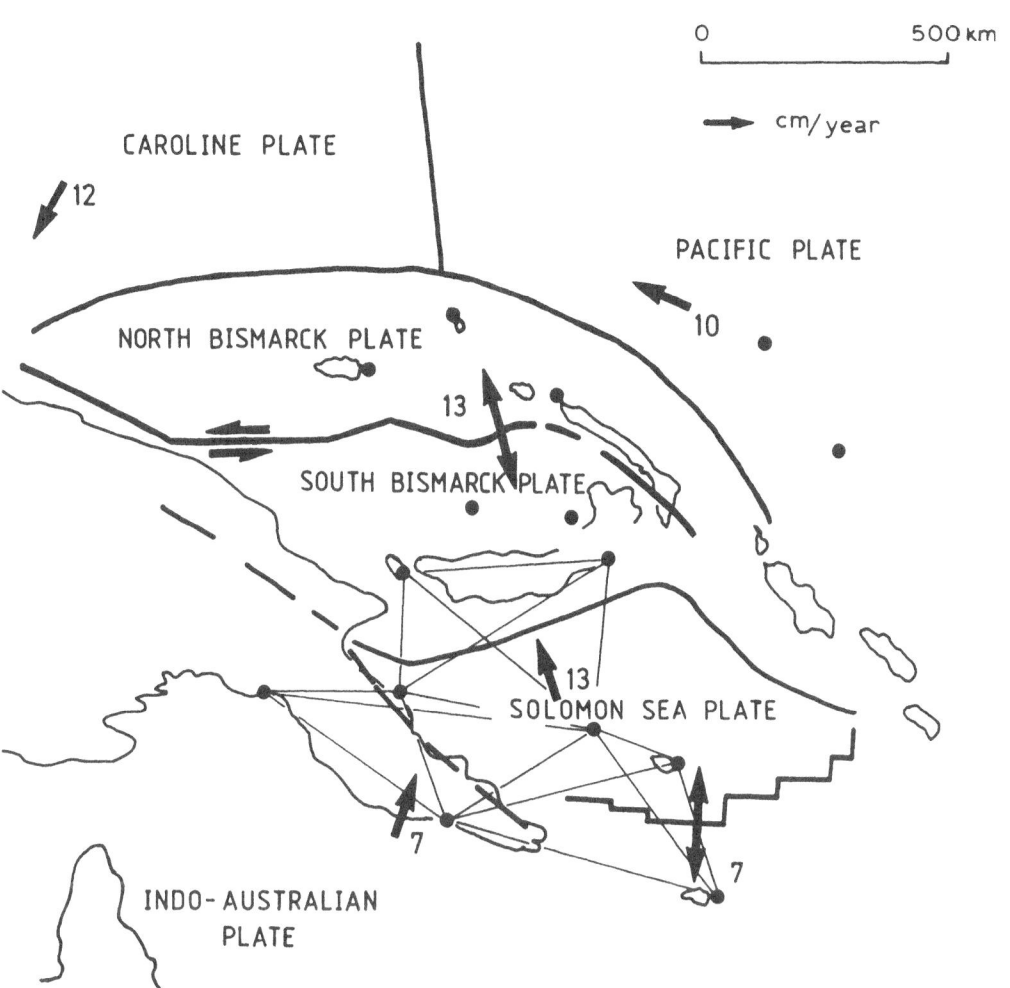

Figure 2: Proposed crustal motion network to be surveyed in 1990 with GPS. Stations of the 1981 Doppler crustal plate survey are shown by solid circles. Arrows indicate the general direction of plate tectonic motion.

Bismarck and Solomon Sea plate in the direction of this baseline should have accumulated to about 120 cm since the Doppler survey was performed, if we observe the network in 1990 and assume that the motion is uniform. As the Doppler measurements are accurate to about 30 cm, we should be able to determine subduction after the first measurement campaign. This line also straddles the Trobriand Trough. However, subduction rates along this trough and the New Britain Trench are likely to vary laterally (B. Taylor, 1988, personal communication) and as long as the variations are not coincident, the difference in the rates obtained from the Woodlark Is. and Trobriand Is. stations to those located on New Britain, should allow the contribution from each subduction zone to be separated. The baseline between Woodlark Is. and Misima Is. is about 180 km long. If we could attain a relative accuracy of 0.1 ppm, the typical baseline uncertainty is 2 cm. Spreading rates in the Woodlark Basin are about 7 cm/a and tectonic signal should have accumulated to about 60 cm by 1990. Thus, we should be able to determine this rate of spreading within a small fractional uncertainty.

Figure 2 also shows the configuration if, as is likely, the network is observed with a suite of four GPS receivers. In this case, it would be expedient to divide the network into three subnetworks of four stations each. A central station, Losuia, could be continuously occupied during the campaign. This station would serve to reduce baseline length, which would be beneficial for phase ambiguity resolution (see below). A considerable amount of redundancy should, moreover, be built into the survey. This redundancy would be used to increase accuracy, thereby strengthening our conclusions about crustal motion determined from this initial project and from future repeat GPS surveys.

6. ANTICIPATED GPS PERFORMANCE

Satellite and receiver oscillator errors are the major error sources which degrade GPS baseline accuracy. These error sources can be eliminated by special processing strategies such as differencing of the GPS phase observables (King et al., 1987). Other significant error sources are: (1) variations in ionospheric refraction; (2) uncertainties in the refractive index of the troposphere largely due to variations in water vapour content; (3) errors in the positions of the GPS satellites; (4) antenna multipathing; and (5) ambiguities in resolving the carrier phase of the received signals.

Ionospheric Refraction

Variations in ionospheric refraction are effectively eliminated by using two frequencies (1.228 MHz and 1.575 MHz) and by exploiting the frequency dependence of the GPS observables. Nevertheless, since the region is close to the "equatorial anomaly" and since the measurements, if performed in 1990, would be made around the solar maximum, ionospheric noise may be a problem, despite the use of dual-frequency receivers. The highest ionospheric refraction effects in the world occur in the equatorial regions. Studies conducted by Klobuchar (1983) indicate extremely large day-to-day variability of the ionosphere during the afternoon and evening hours in these regions and it would be best to avoid these times by performing the observations at night, as much as possible. Onsite preprocessing would also allow identification of such problems in time to repeat measurements during the measurement campaign, if necessary.

Tropospheric Refraction

Tropospheric refraction effects may be suppressed in a number of ways. Tralli et al. (1988) describe the techniques which are currently in use with high-precision GPS. One way is to determine the tropospheric delay directly from measurements made with a water vapour radiometer (WVR) and to subsequently calibrate the observations. However, WVRs are presently very scarce and it may be too optimistic to consider their use in an initial experiment. For a network of sites in the Gulf of California, where tropospheric conditions are similar to those that we expect in Papua New Guinea, Tralli et al. (1988) obtained horizontal baseline component accuracy of about 0.1 ppm over distances of 450-650 km using WVR calibration of the GPS observables and stochastic estimation of the residual zenith delay. Similar accuracies were obtained from stochastic estimation without prior calibration of the tropospheric delay. Thus, stochastic estimation schemes may be an attractive alternative to WVR calibration at humid sites as they occur in Papua New Guinea. Yet another technique is to take surface measurements of pressure, humidity and temperature, use these to calibrate the GPS data and to subsequently estimate the residual tropospheric zenith delay along with the other parameters of the problem. While Tralli et al. did not achieve a great deal of success with this technique, it has considerable merit and it would be worth trying out in Papua New Guinea.

GPS Orbits

The effects of uncertainties in the positions of the satellites on baseline accuracy may be reduced by tracking the satellites with an independent suite of GPS receivers positioned at known ground locations (eg VLBI or satellite laser ranging sites). Lichten and Border (1987) used this fiducial concept to explore different strategies for high-precision orbit determination with data from 1985 field experiments conducted in the United States. Their most successful strategy employs multi-day arcs and incorporates fine tuning of spacecraft solar pressure coefficients and stochastic zenith tropospheric delays using GPS data. Using these orbits on independently observed baselines 250-1 300 km long produces a baseline accuracy of better than 0.1 ppm for all components (Bertiger and Lichten, 1987). Results from other geographical locations using the free-network approach indicate similar accuracies (Beutler et al., 1987; Rothacher et al., 1987). This implies that this error source would not be an important problem for the baselines shown in Figure 2.

Antenna Multipath

Antenna multipath is a site-dependent phenomenon. It is a delay noise source that depends not only on the geometry of the observing site, but also on the material and texture of the surfaces in the area surrounding the antenna (Tranquilla, 1986). While Georgiadou and Kleusberg (1987) succeeded in modelling multipath from a flat and horizontal surface, it is unlikely that all possible geometric configurations which can produce multipath effects will ever be modelled satisfactorily. Multipath can be reduced by selecting a suitable location for the GPS antenna and by surrounding the antenna with absorbing material to prevent ground reflections from reaching the phase centre (Evans, 1986).

Ambiguity Resolution

The remaining major error source, phase ambiguity resolution, can be eliminated if the other error sources are reduced to several centimetres or less. The problem arises in distinguishing between adjacent cycles in the carrier signal in the presence of path delay noise, which increases in difficulty with increased baseline length (Bender and Larden, 1985; Melbourne, 1985). Due to the predominantly north-south velocity distribution of the present GPS constellation with respect to stations on the earth's surface, carrier phase bias parameters are correlated with the longitudes of the estimated station positions. Lichten and Border (1987) found that the addition of pseudorange data to the precise but ambiguous carrier phase measurements helped substantially in overcoming this problem, if the phase ambiguities are estimated. Some success has also been reported with "bias fixing" over baselines up to about 1 000 km long (Bock et al., 1985, 1986; Dong and Bock, 1989). Bias fixing refers to constraining the phase biases to integer values, and effectively removing the biases as parameters from the solution. Blewitt (1988), on the other hand, believes that it is generally a poor strategy to indiscriminantly fix every bias to the nearest integer value, and prefers to calculate the cumulative probability that all the widelane and ionosphere-free bias fixes have been corrected, and to subsequently fix another bias only if the cumulative probability exceeds a certain confidence limit. There are, moreover, two major factors which strengthen ambiguity resolution for networks when compared to individual baselines: (1) ambiguities for longer baselines are often resolved as the linear combination of ambiguities for shorter baselines; and (2) all the ambiguities are correlated. Thus, by first resolving the best determined ambiguities, the solutions for the remaining ambiguities are strengthened. This means that some of the baselines of the crustal motion network should be short, of the order of 100 km, if longer baselines are to be measured.

7. CONCLUDING REMARKS

We do not know if the limit of accuracy for GPS measurements would improve beyond 0.1 ppm. The use of WVRs for the wet path correction at a few sites, improvements in baseline reduction software, refined field procedures, and the availability of additional satellites may increase accuracy. On the other hand, possible large and variable wet path delays in tropical island climates and a less than optimum satellite configuration over Papua New Guinea during preliminary stages of the project, could lead to degradation of accuracy. Taking these considerations into account, we believe it reasonable to expect accuracies similar to those achieved in the Gulf of California, that is, at the 0.1 ppm level for all baseline components. We also believe that the rapid tectonic movements characteristic of this region, in combination with short and moderate baseline lengths imply some of the highest signal-to-noise ratios achievable in this kind of experiment anywhere in the world.

ACKNOWLEDGEMENTS

The contributions made to this paper by Fritz Brunner, Ken Hurst, Kurt Lambeck and Peter Morgan are gratefully acknowledged.

References

Abbot, R.I., Bock, Y., Counselman, C.C., King, R.W., Gourevitch, S.A. and Rosen, B.J., 1985, Interferometric determination of GPS satellite orbits, in C.C. Goad ed., *Proc First Int. Symp. on Precise Pos. with GPS*, May 1985, Rockville, Maryland, NOAA, 447-456.

Angus-Leppan, P.V., Allman, J.S. and Sloane, B., 1983, Crustal movement from satellite observations in the Australian region, *Tectonophys., 97,* 87-93.

Bender, P.L. and Larden, D.R., 1985, GPS carrier phase ambiguity resolution over long baselines, *Proc First Int. Symp. on Precise Pos. with GPS,* May 1985, Rockville, Maryland, NOAA, 357-362.

Bertiger, W. and Lichten, S., 1987, Demonstration of 5-20 parts per billion repeatability for continental baselines estimated with multi-say GPS orbits, *EOS Trans. AGU, 68,* 1238.

Beutler, G., Bauersima, I., Gurtner, W., Schildknecht, T., Mader, G.L. and Abell, M.D., 1987, Evaluation of the 1984 Alaska campaign with the Bernese GPS software, *J. Geophys. Res., 92,* 1295-1303.

Bevis, M. and Bock, Y., 1989, 'APEX', densification of GPS global tracking network in Asia and the Pacific, IAG Comm. VIII, Int. Coord. Space Techn. Geod. Geodynamics (CSTG), GPS Subcomm., *GPS Bull., 2(1),* 1.

Blewitt, G., 1988, Carrier phase ambiguity resolution for the Global Positioning System applied to geodetic baselines up to 2 000 km, *J. Geophys. Res.,* in press.

Bock, Y., Counselman, C.C., Gourevitch, S.A. and King, R.W., 1985, Establishment of three-dimensional control by interferometry with the Global Positioning System, *J. Geophys. Res., 90,* 7689-7703.

Bock, Y., Gourevitch, S.A., Counselman, C.C., King, R.W. and Abbot, R.I., 1986, Interferometric analysis of GPS observations, *Manuscr. Geod., 11,* 282-288.

Chappell, J., 1974, Geology of coral terraces, Huon Peninsula, New Guinea - a study of Quaternary tectonic movements and sealevel changes, *Geol. Soc. Am. Bull., 85,* 553-570.

Connelly, J.B., 1976, Tectonic development of the Bismarck Sea based on gravity and magnetic modelling, *Geophys. J. Roy. astron. Soc., 47,* 23-40.

Cook, D.P. and Murphy, B., 1974, Crustal movement survey: Markham Valley, Papua New Guinea, 1973, *Div. Natl. Mapping Tech. Rep., 18.*

Cooper, P. and Taylor, B., 1987, The spatial distribution of earthquakes, focal mechanisms, and subducted lithosphere in the Solomon Islands, in B. Taylor and N.F. Exon, eds., 1987, *Marine Geology, Geophysics and Geochemistry of the Woodlark Basin-Solomon Islands,* Circum-Pacific Counc. Energy Miner. Res. Earth Sci. Ser., 7, Houston, Tex., 67-88.

Curtis, J.W., 1973, Plate tectonics of Papua New Guinea - Solomon Islands region, *J. Geol. Soc. Aust., 20,* 21-36.

D'Addario, G.W., Dow, D.B. and Swoboda, R., 1976, Geology of Papua New Guinea 1:2 500 000, Bur. Min. Res., Australia.

Davidson, J.M., Skrumeda, L.L., Tralli, D.M. and Thornton, C.L., 1987, Comparison of March 1985 and November 1985 GPS-based measurements of the Mojave-Ownes Valley-Hat Creek baselines, *EOS Trans. AGU, 68,* 283.

Dong, D. and Bock, Y., 1989, GPS network analysis with phase ambiguity resolution applied to crustal deformation studies in California, *J. Geophys. Res., 94,* 3949-3966.

Evans, A.G., 1986, Comparison of GPS pseudorange and biased Doppler range measurements to demonstrate signal multipath effects, *Proc. Fourth Int. Symp. Sat. Pos.,* 28 April - 2 May, 1985, 573-587.

Falconer, R.H.K., 1973, Indian-Pacific rotation pole determined from earthquake epicentres, *Nature, 243,* 97-99.

Georgiadou, Y. and Kleusberg, A., 1987, Ionospheric refraction and multipath effects in GPS carrier phase observations, *Symposium U3, Impact of GPS on Geophysics,* IUGG XIX General Assembly, Vancouver, 9-22 August, 1987, unpublished.

Gornitz, V.S., Lebedeff, S. and Hansen, J., 1982, Global sealevel trend in the past century, *Science, 215,* 1611-1614.

Hamilton, W., 1979, Tectonics of the Indonesian region, *Geol. Surv. Prof. Paper 1078,* US Gov. Printing Office, Washington, D.C.

Hansen, J., Johnson, D., Lacis, A., Lebedeff, S., Lee, P., Rind, D. and Russell, G., 1981, Climate impact of increasing atmospheric carbon dioxide, *Science, 213,* 957-966.

Honza, E., Davies, H.L., Keene, J. and Tiffin, D.L., 1987, Plate boundaries and evolution of the Solomon Sea region, *Geo-Marine Lett., 7,* 161-168.

Jaques, A.L. and Robinson, G.P., 1978, The continent/island arc collision in northern Papua New Guinea, *BMR J. Aust. Geol. Geophys., 2,* 289-303.

Johnson, T. and Molnar, P., 1972, Focal mechanisms and plate tectonics of the southwest Pacific, *J. Geophys. Res., 77,* 5000-5032.

Johnson, R.W., 1979, Geotectonics and volcanism in Papua New Guinea: a review of the late Cainozoic, *BMR J. Austr. Geol. Geophys., 4,* 181-207.

Johnson, R.W., 1987, Delayed partial melting of subduction - modified magma sources in western Melanesia: new results from the late Cainozoic, *Proc. Pacific Rim Congr.,* Gold Coast, Australia, August, 1987, 211-214.

King, R.W., Masters, E.G., Rizos, C., Stolz, A. and Collins, J., *Surveying with the Global Positioning System - GPS,* Ferd. Dümmler, Bonn.

Klobuchar, J.A., 1983, Ionospheric effects on earth-space propagation, *AFGRL-TR-84-0004 Environmental Res. Papers, 866,* AFGRL, Hanscomb AFB, Mass.

Krause, D.C., 1973, Crustal plates of the Bismarck and Solomon Seas, in R. Fraser ed., *Oceanography of the South Pacific 1972*, New Zealand Nat. Comm. UNESCO, Wellington, 271-280.

Lichten, S.M. and Border, J.S., 1987, Strategies for high-precision Global Positioning system orbit determination, *J. Geophys. Res., 92,* 12751-12762.

Lock, J., Davies, H.L., Tiffin, D.L., Murakami, F. and Kisimoto, K., 1987, The Trobriand subduction system in the western Solomon Sea, *Geo-Marine Lett. 7,* 129-134.

Marlow, M.S., Exon, N.F. and Tiffin, D.L., 1988, Lava flows, sediment deformation and arc rotation in the Manus forearc, Northern Papua New Guinea, in preparation.

Melbourne, W.G., 1985, A case for ranging in GPS based geodesy, in C.C. Goad ed., *Proc First Int. Symp. on Precise Pos. with GPS,* May 1985, Rockville, Maryland, NOAA, 373-386.

Minster, J.B. and Jordan, T., 1978, Present-day plate motions, *J. Geophys. Res., 83,* 5331-5354.

Molnar, P., Atwater, T., Mammerickx, J. and Smith, S.M., 1975, Magnetic anomalies, bathymetry and tectonic evolution of the South Pacific since the late Cretaceous, *Geophys. J. Roy. astr. Soc., 40,* 383-420.

Mori, J., 1988, The New Ireland earthquake of July 3, 1985 and associated seismicity near the Pacific-Solomon Sea-Bismarck Sea triple junction, *Phys. Earth Planet. Int.,* submitted.

Rothacher, M., Beutler, G., Gurtner, W., Mader, G.L. and Abell, M.D., 1987, Results of the Alaska GPS campaign: comparison of the 1984 GPS solution with VLBI, *EOS Trans. AGU, 68,* 283.

Sloane, B.J. and Steed, J.B., 1976a, Crustal movement survey: Markham Valley, Papua New Guinea 1975, *Div. Natl. Mapping Tech. Rep., 23.*

Sloane, B.J. and Steed, J.B., 1976b, Crustal movement survey: St. Georges Channel, Papua New Guinea 1975, *Div. Natl. Mapping Tech. Rep., 24.*

Stolz, A., Masters, E.G. and Rizos, C., 1984, Determination of GPS satellite orbits for geodesy in Australia, *Austr. J. Geod. Photogram Surv., 40,* 41-52.

Taylor, B., 1975, The tectonics of the Bismarck Sea region, B.Sc., thesis, University of Sydney, unpublished.

Taylor, B., 1979, The Bismarck Sea: evolution of a back-arc basin, *Geology, 7,* 171-174.

Taylor, B., 1987, A geophysical survey of the Woodlark-Solomons region, in B. Taylor and N.F. Exon, eds., 1987, *Marine Geology, Geophysics and Geochemistry of the Woodlark Basin-Solomon Islands,* Circum-Pacific Counc. Energy Miner. Res. Earth Sci. Ser., 7, Houston, Tex., 25-48.

Taylor, B., Sinton, J., Liu, 1. and Crook, K., 1986, Extensional transform zone, Manus back-arc basin, *EOS Trans. AGU, 67,* 377.

Tingey, R.J. and Grainger, D.J., 1976, Markham, Papua New Guinea, 1: 250 000 geological map and explanatory notes, Geol. Surv. Papua New Guinea, Sheet SB/-10.

Tralli, D.M., Dixon, T.H. and Stephens, S.A., 1988, The effect of wet tropospheric path delays on estimation of geodetic baselines in the Gulf of California using the Global Positioning System, *J. Geophys. Res., 93,* 6545-6558.

Tranquilla, J.M., 1986, Multipath and imaging problems in GPS receiver antennas, *Proc. Fourth Int. Symp. Sat. Pos.,* 28 April - 2 May, 1985, 557-571.

Weissel, J.K. and Anderson, R.N., 1978, Is there a Caroline plate?, *Earth Planet. Sci. Lett., 41,* 143-158.

Weissel, J.K., Taylor, B. and Karner, G.D., 1982, The opening of the Woodlark Basin, subduction of the Woodlark spreading system and the evolution of northern Melanesia since mid-Pliocene times, *Tectonophys., 87,* 253-277.

The Determination of Present-Day Tectonic Motions from Laser Ranging to LAGEOS

D.E. Smith & R. Kolenkiewicz
NASA/Goddard Space Flight Center, Greenbelt Md.
U.S.A.

P.J. Dunn, M.H. Torrance, S.M. Klosko, J.W. Robbins &
R.G. Williamson
ST Systems Corp., Lanham Md.
U.S.A.

E.C. Pavlis & N.B. Douglas
University of Maryland, College Park Md.
U.S.A.

S.K. Fricke
RMS Technologies Inc., Landover Md.
U.S.A.

ABSTRACT:

Over twelve years of laser ranging to the LAGEOS spacecraft have enabled the motions of the Earth's crust to be determined at approximately twenty laser tracking sites around the world. These motions show the surface of the Earth to be moving in general accord with the theory of plate tectonics and to deviate from the principle of rigid plates only in regions near plate boundaries. In western North America, along the Pacific and North America Plate boundary, the motions of the individual sites move considerably less than the full plate motion, primarily since motion is spread over a series of faults across a relatively broad boundary zone. Between Quincy (in northern California) and Monument Peak (40 km east of San Diego) the relative motion determined in our solution is only 26±2 mm/yr compared to the AM0-2) of Minster & Jordan (1978). In Australia, the relative motion of Yaragadee with respect to Hawaii is, from our solution, -89±2 mm/yr compared to the AM0-2 predicted value of -103 mm/yr. The motion between the South American site at Arequipa, Peru and Greenbelt on the North American Plate, is in close agreement with the geologic model; having only a few mm/yr compression. The motion across the Mid-Atlantic Ridge between Greenbelt and Wettzell (on the Eurasian Plate) is, from our solution, determined to be 14±2 mm/yr compared to an AM0-2 predicted rate of 21 mm/yr. The relative motion of Hawaii and Arequipa is 80±3 mm/yr from our solution compared to the geologically predicted 66 mm/yr.

INTRODUCTION

Global-scale contemporary tectonic motions were largely unobservable before the advent of space geodetic techniques. Several global plate motion models have been derived, however, using seismological, geological and geomagnetic results representing average motion for the last few million years. [Minster and Jordan, 1974, 1978; Chase, 1972, 1978; and DeMets et al., 1989]. There remain however, unresolved questions about the accuracy of plate motion models and their usefulness in describing present-day motions at diffuse plate boundaries. Classical, ground-based, geodetic measurements have provided localized tectonic information in the form of regional strain (over distances of tens of kilometers) and deformation results. In the late 1960's, two evolving space geodetic techniques, Satellite Laser Ranging (SLR) and Very Long Baseline Interferometry (VLBI), were shown to have the potential for detecting tectonic motions at the level of a centimeter or better per year. Considerable progress continues to be made in both SLR and VLBI technologies as well as other geodetic positioning systems being developed (e.g. GPS).

The most comprehensive publication of SLR results to date can be found in the Journal of Geophysical Research LAGEOS (Laser Geodynamic Satellite) special issue [Cohen and Smith, 1985]. Current results have been reported by various groups at meetings of the National Aeronautics and Space Administration's (NASA) Crustal Dynamics Project and at the American Geophysical Union meetings [e.g., Smith et al., 1989a and Schutz et al., 1989]. More recent geophysical results from VLBI measurements have been reported by Clark et al. [1987] and Ma et al. [1989]. Integration of space geodetic data into regional tectonic studies have been reported by Minster and Jordan [1987], Kroger et al. [1987], and Jordan and Minster [1988], among others.

This paper describes a portion of the results from analysis of data from the LAGEOS Satellite by the Geodynamics Branch of NASA's Goddard Space Flight Center (GSFC). Laser data used in this study span from 1979 to midway into 1988 and were collected by the Goddard Laser Tracking Network (GLTN) stations, the Smithsonian Astrophysical Observatory (SAO) stations and numerous international stations. Figure 1 is a map showing their locations. This present solution, SL7.1, is GSFC's most recent solution for the motions of the laser tracking sites and incorporates the latest force, reference system and measurement models. The GEM-T1 gravity model of Marsh et al. [1988] and the GEM-T1 tidal model of Christodoulidis et al. [1988] have been used. A more detailed discussion of the analysis can be found in Smith et al., [1989b] and the previous GSFC tectonic/SLR solution of Christodoulidis et al., [1985].

METHODOLOGY

The estimation of useful geophysical and geodetic information from SLR observations relies on sufficient observational geometries (i.e. global station coverage and sky coverage for measurements taken at an individual station) and appropriate modeling of the orbit dynamics. In our solution, the data are related in time through the equations of motion which are used to recover the evolution of LAGEOS orbit. Since LAGEOS is a passive satellite, we do not need to estimate time-variant satellite dependent processes such as fuel consumption and moments of inertia. It's spherical shape and high density also help to simplify the kinematics of the problem. Known forces which perturb the spacecraft are accurately modeled or estimated. Earth orientation parameters are implicitly referenced to a system based on the LAGEOS orbit and the estimated station positions. Table 1 lists the nominal models and constants that enter into the reduction of the observations. We have avoided repeating numbers which are implicitly embedded in standard models which we've adopted. To facilitate intercomparison and distribution, we have tried, wherever possible, to follow the internationally accepted MERIT Standards [Melbourne et al., 1983].

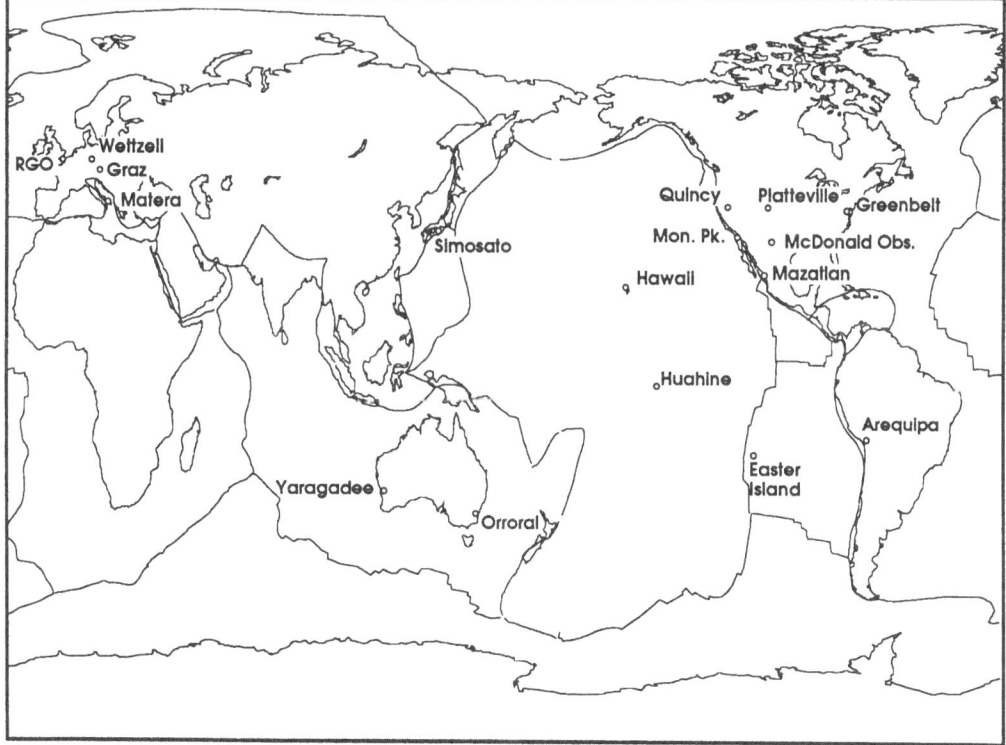

Fig.1. Satellite Laser Ranging sites used in this study.

Table 1. Description of SL7.1

Force Model
- Gravity: GEM-T1
- Earth & ocean tides: GEM-T1
- Luni-solar and planetary gravitational perturbations (Mercury through Neptune)
- Solar radiation pressure includes lunar eclipse model

Analysis Techniques
- Along track acceleration estimated every 15 days
- Solar radiation coefficient estimated every 15 days
- Capable of estimating station dependent range & timing biases

Data Reduction
- Normal points (2 minute bins): Herstmonceaux recommendations
- Improved data editing on a pass by pass basis
- Observation history extends from May 1976 to June 1988

Reference System
- Tectonic: Minster & Jordan AM0-2
- Earth rotation and orientation every 5 days
- Wahr's nutation series
- JPL DE-200 planetary ephemerides: J2000 reference system

Our dynamical method is based on the differential improvement of an *a priori* model. It utilizes the observed LAGEOS range. measurements and compares them with synthetic data generated by the model. The observation equations, which relate the observed range measurements to the computed ones, and the corrections to the a priori values of the model parameters, involve those models either explicitly or implicitly. Of particular importance is the implementation of the *a priori* tectonic plate motion model. Since laser range measurements contain no direction information, our solution for site velocities requires some external datum definition. We accomplished this in the orbit and stations position estimation process by defining the motion of two sites which define our Conventional Terrestrial Reference System (CTRS): latitude and longitude of Greenbelt, Maryland and latitude only of Maui, Hawaii. The AM0-2 model of Minster and Jordan [1978], with no net lithospheric rotation, has been used as the *a priori* model for SL7.1 (denoted here as M/J). This absolute motion model has been widely used and its implied relative motion for the constrained sites is nearly identical to the SLR observed motion.

Figure 2 is an overview of the SL7.1 solution design; it shows the flow of the process from the acquisition of LAGEOS laser return pulses, through all of the successive steps required to compute a reliable estimate of horizontal site velocities. This process begins with the compression of the full-rate laser data into two-minute intervals (normal points) with the use of the GEODYN orbit determination and parameter estimation software system developed at GSFC [Putney, 1976; Martin et al., 1987]. Typical data compression ratios are 1:100 to 1:1000, and depend upon the data rate of the laser system. Extensive testing of the normal point process has been made to insure that no significant loss of geodetic or orbit information occurs [Torrence et al., 1984]. These refined tracking data are then partitioned into time spans of approximately 30 days constituting an *arc*. Each of these arcs is separately processed, evaluated, and qualified. The arc length has been chosen for its dynamic strength (~180 continuous satellite revolutions). The data usually fit to within a few centimeters over an arc interval, which supports the adequacy of this choice of data division. Anomalous data due to mechanical or systematic errors are either edited out, or resolved and corrected prior to and during successive iterations of the individual arcs.

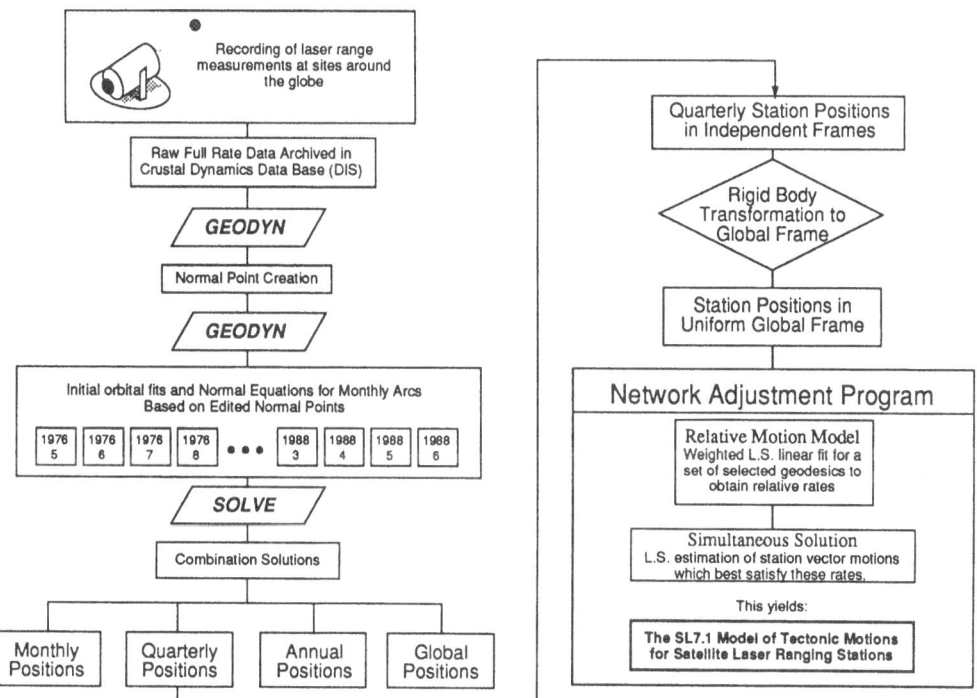

Fig. 2. Diagram illustrating SL7.1 solution design. GEODYN and SOLVE are software packages described in the text.

Once the data validation and qualification steps are completed, GEODYN is used to generate normal equations for all of the parameters to be estimated. Next, these normal equations are inverted to yield a selected subset of parameters to further assess data performance in local monthly solutions. An iterative procedure is then used to verify that all data problems have been resolved and that the normal equations are based upon proven observations. Next, the monthly normal equations are then combined into groups of three months (calendar quarters) using a GSFC software package, SOLVE, which performs the matrix inversion and estimation of the *quarterly* solution parameters. These quarterly solutions yield averaged three-dimensional station coordinates, which are tied dynamically to the geocenter by LAGEOS' strong sensitivity to the Earth's long-wavelength gravitational field.

Earth orientation and orbital arc parameters are simultaneously estimated within each quarterly solution. Figures 3a and 3b are examples of how the dynamical aspects of the Earth's rotation do not remain fixed in time or space, but instead exhibit considerable variation. Figure 3a is a plot of the location of the spin axis of the Earth as a function of time. While this variation is dominated by the 14-month Chandler Wobble, there are also annual and shorter period signals which cause additional variation and cannot be adequately predicted from models. Figure 3b shows the length of day as measured by LAGEOS laser ranging, from launch through 1988. It reflects a secular variation which is different from the overall trend. The Earth's rotation is known to be slowing, due to the decreasing angular momentum of the Earth-Moon system; this should result in an increasing length of day. However, current measurements show a decreasing length of day, which has been related to decadal fluctuation in the rotation of the earth. These variations are also not entirely predictable, although there are strong annual signals due to the exchange of atmospheric angular momentum. There are also correlations between the length of day and certain atmospheric processes, such as the El Niño Southern Oscillation. The large peak on figure 3b during 1982-83 corresponds well to a maximum El Niño the following year.

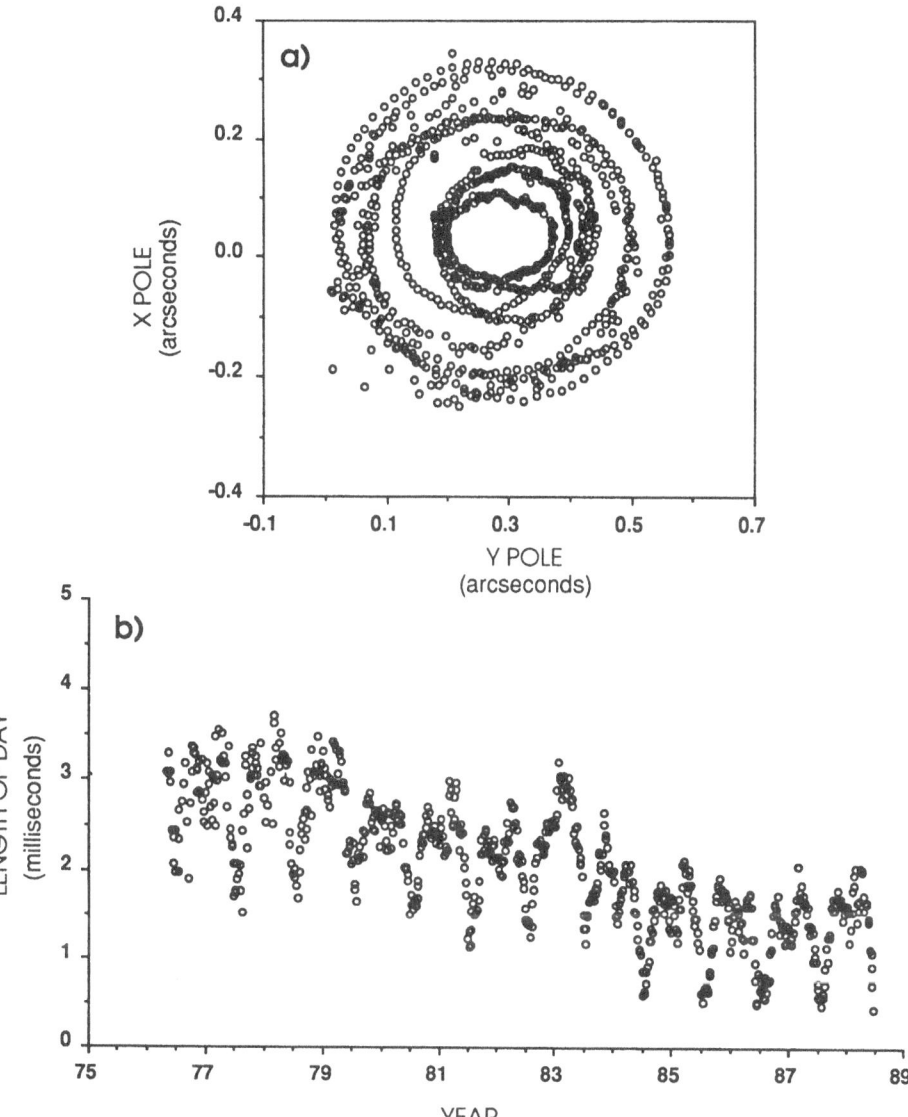

Fig.3. (a) Polar motion recovered from the analysis of SLR data, May 1976 to June 1988. (b) Recovered length of day variations obtained by forward differencing of the estimated (UT1-TAI) values over the same period.

The decreasing monthly orbital RMS of fit for LAGEOS is shown in figure 4a. In the early years, between 1976 and 1978, this RMS was around 30 cm, with a fair amount of scatter. For this reason, we have elected to omit data from this time period from further consideration in the present work. As tracking accuracy and modeling capabilities improved and as more stations were added to the network, the RMS and its scatter have decreased considerably. For the last few years this RMS has been consistently under 5 cm. Figure 4b reflects a similar pattern for monthly GM values. Before 1980, considerable variations and large uncertainties prevailed, but with increased data, the value of GM has stabilized around a mean of 398600.4408 km^{3}s^{-2}. This suggests that the original variation was not a real phenomena, but is rather an artifact of data quality and quantity.

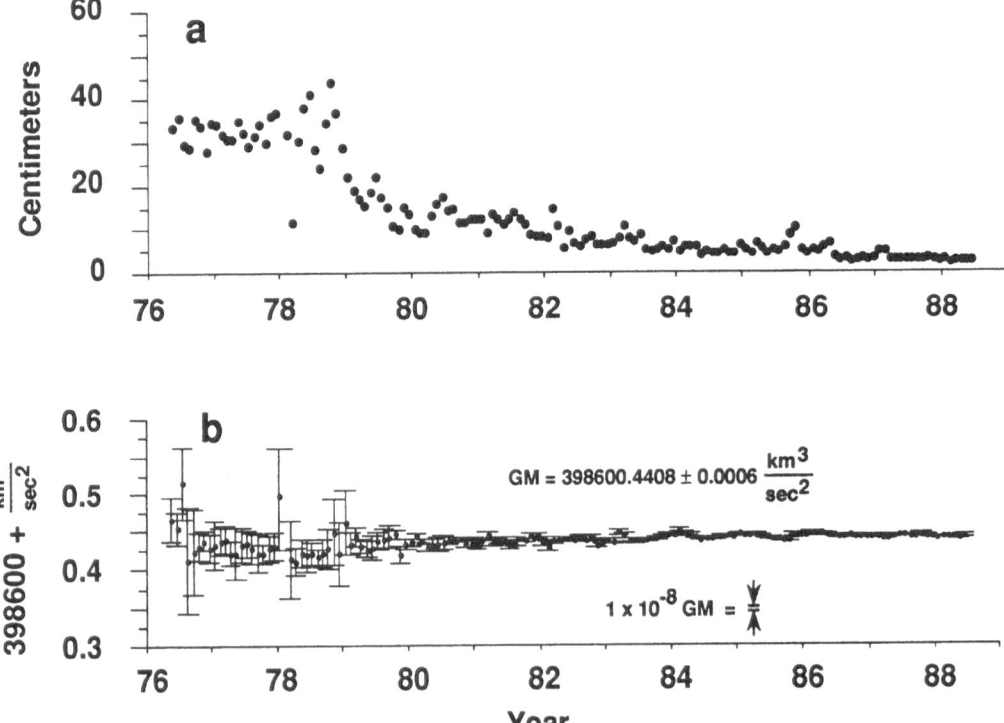

Fig. 4. (a) RMS of orbital fit for each monthly arc of LAGEOS data. (b) Values of the product of the mass of the Earth and the gravitational constant (GM) determined in each monthly arc.

Once the three-dimensional, quarterly station positions are determined, additional processing is required to remove reference frame instabilities. These discontinuities are due largely to the nature of the network geometry and the variable participation of sites. These reference frame "jitters" are reduced in our analysis by means of rigid body transformations (three translations and three rotations), which bring all of the quarterly geodetic coordinates into a common reference frame. Intersite baseline distances are strictly preserved, but the network is slightly re-positioned with respect to a more stable definition of the center of mass.

These "transformed" quarterly solutions are then used to solve for geodesic rates of change between pairs of stations having made observations during the same calender quarter. Examples of two such rates are shown in Figures 5a and 5b. These plots show the time evolution of two geodesic distances radiating from Yaragadee, based upon quarterly geodesic length. These geodesic distance rates are used to describe the relative horizontal changes in the positions of the sites on the Earth's reference ellipsoid. Thus the computed rates are independent of station vertical motions and their height uncertainties. These inter-site geodesic changes are then used to derive SLR site velocities in a global network adjustment program. Station motions are estimated within the reference frame in a least squares estimation procedure. For satisfactory closure, the assumed motion vector for each of the reference stations must imply relative motions compatible with those observed directly by SLR. The network adjustment estimates motion for 17 of the stronger stations in the SLR network relative to an AM0-2 based frame realized by two reference stations for which the horizontal motion is considered perfectly known; that is, we adopt AM0-2 to describe the motions of Greenbelt and Maui. In total, there are 170 geodesic distance rates used in the network adjustment to determine the 34 site velocity components.

These velocities are then used to compute the network implied geodesic rates between all sites used in the network adjustment, along with their respective uncertainties determined through error propagation. This increases the amount of information available, because rates can be inferred from this solution regardless of whether there was actual laser tracking at each site during common quarters. However, some caution must be exercised when interpreting these adjusted rates. Not all of the resultant rates have come from equally robust, high quality data, and some weak geometries or time evolutions can occur. Figures 6a and 6b illustrate the geodesic rates (both those recovered in this solution and those predicted by AM0-2) radiating from two sites in Australia to other sites in the SLR network. The projection used in this figure preserves the azimuths for lines radiating from the sites and accurately depicts the plate boundary crossings by the geodesics between each site. The geophysical implications from these figures will be discussed in the next section.

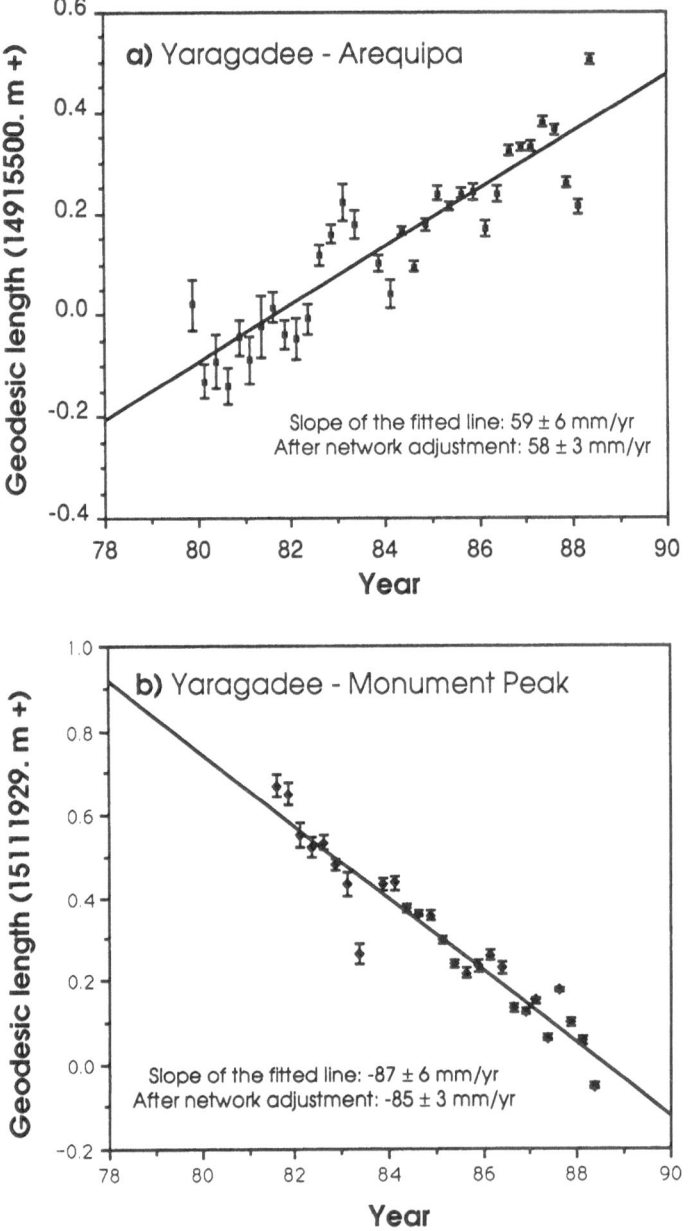

Fig. 5. Time history of geodesic lengths between Yaragadee to Arequipa (a) and Yaragadee to Monument Peak (b). Sigmas quoted for slopes and sigma bars shown are at the 67% confidence level.

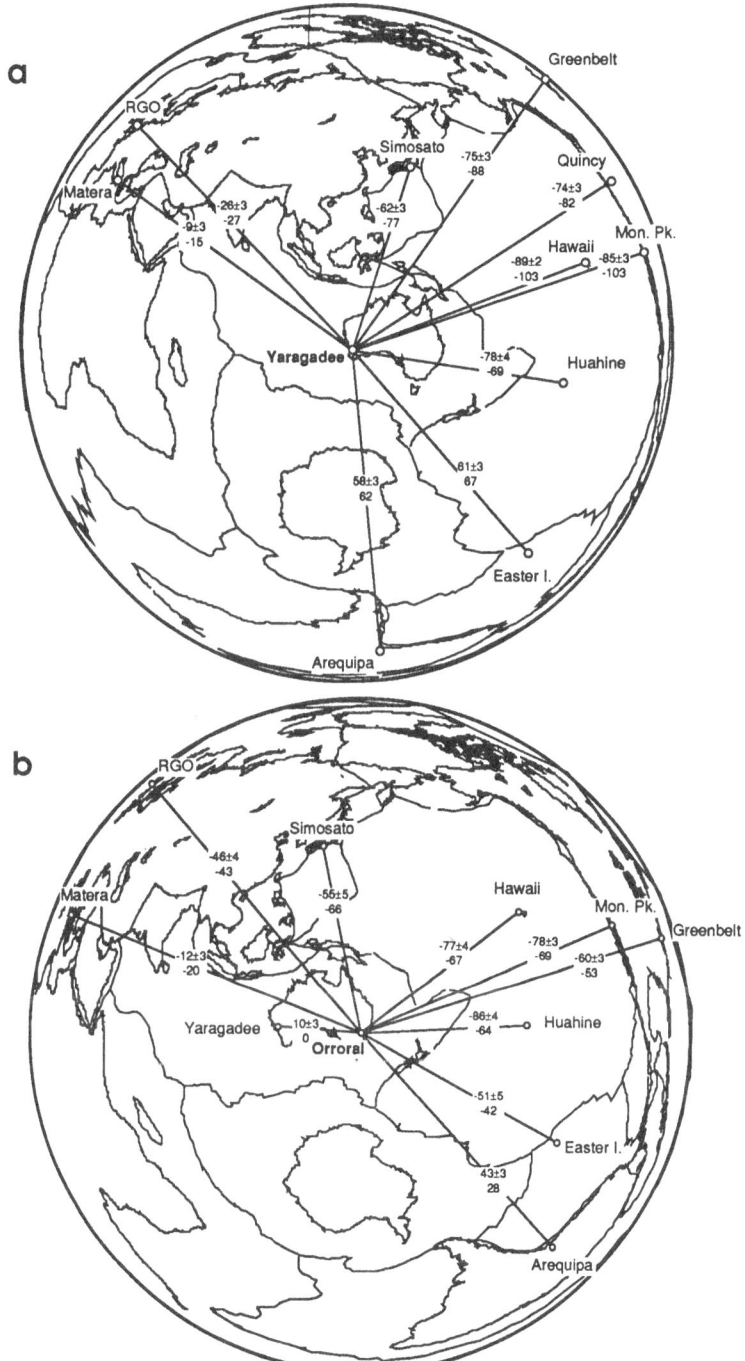

Fig. 6. Selected geodesic rates for lines from Yaragadee (a) and for lines from Orroral (b). On each line, top values are determined from the SLR analysis and are the rates implied by the SLR model (i.e. after network adjustment). Bottom values are those predicted by the Minster & Jordan AM0-2 model. The map is based on a Lambert Azimuthal Equal Area projection whereby azimuths are correctly represented from the central point.

GEOPHYSICAL RESULTS

In this work, we have recovered results from sites located on six major tectonic plates. As shown in figure 1, many SLR sites are located on the North American and Eurasian plates, several on the Indo- Australian and Pacific plates, and a single site on the Nazca, African and South American plates respectively. Although the southern hemispheric portion of the African continent is devoid of site occupations, there are two currently active sites on the northern portion of the African Plate which are not represented here: Helwan, Egypt and Bar Giyyora, Israel. Vector motions estimated for these sites are preliminary and not yet statistically reliable. The rigid-plate tectonic models inadequately describe the complexity found at many tectonic boundaries. They can however, provide kinematic boundary conditions which should be satisfied when all local and regional deformations are accounted for. Since the SL7.1 site velocities are reference frame dependent, the intersite geodesic distance variations provide a frame-invariant quantity that is more suitable for intercomparisons with geologic models.

The laser sites having reliable SL7.1 deduced site velocities which are centrally located on their tectonic plates are listed in figure 7. Also shown is the correlation of the geodesic interplate rates for those 10 stations obtained from our LAGEOS analysis and those predicted using the AM0-2 model. The slope and correlation between SL7.1 and AM0-2 rates are 0.99873 and 0.948 respectively. For these SLR sites which are located in the stable plate interiors, the agreement between SLR observations over the last decade and the linear rates predicted from million year averages developed from magnetic anomaly profiles, transform fault azimuths and Earthquake slip vectors, is good. Similar results have been obtained for the NUVEL-1 model. This suggests that geologic models, despite their idealized assumption of rigid-plate behavior, are quite reliable for describing broad scale tectonic motion.

On the Pacific Plate, Maui, Monument Peak and Huahine have provided sufficiently long, accurate data sets for their inclusion here. Since Maui is one of the reference stations and has its motion constrained, Monument Peak and Huahine remain as freely adjusting Pacific stations. Easter Island lies upon the Nazca plate close to the East Pacific Rise. While this is the dominant geologic feature in this region, there is complicated tectonic activity involving numerous microplates believed to be present along the boundary between the Pacific and Nazca plates. Figure 8 shows several baselines around the Pacific Basin, and their associated SLR and AM0-2 modeled geodesic rates. The uncertainties quoted for the SLR modeled rates are at the 67% confidence level. Monument Peak exhibits slightly aberrant motion, very likely due to its proximity to the San Andreas Fault System.

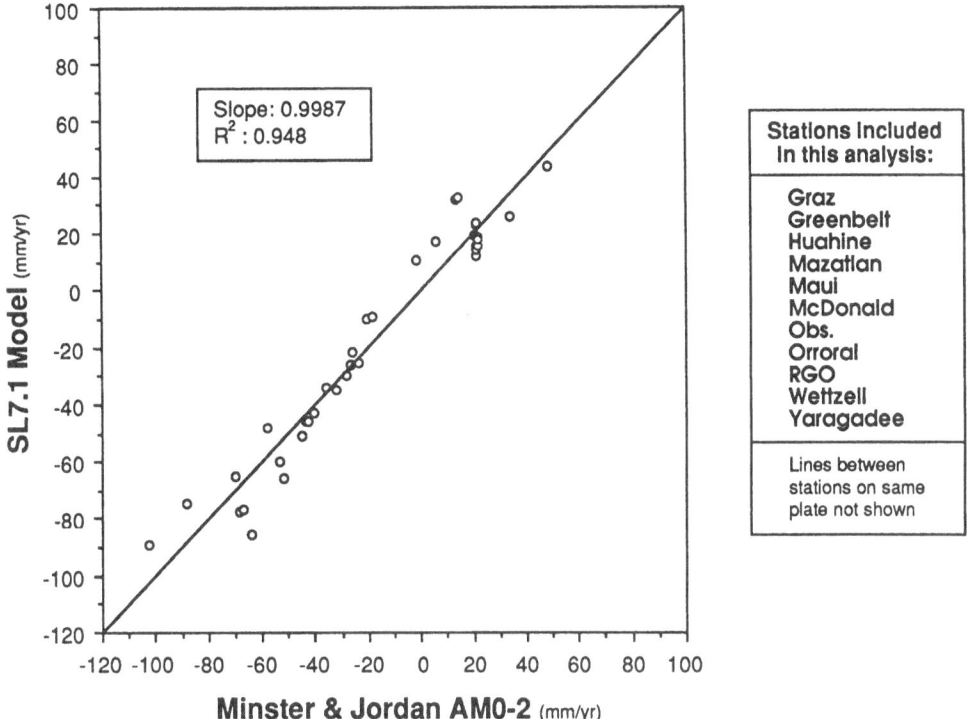

Fig. 7. Correlation between SLR model and Minster & Jordan AM0-2 model of geodesic rates for tracking sites located deep within plate interiors. Slope of nearly 1 suggests very good agreement between the SLR model which is based on 11 years of satellite tracking and the AM0-2 geologic model based on million year averages.

The fastest rates of motion in the entire network are observed between the Nazca and Pacific Plates. Although some of these results are from only a few years of data, the results we have obtained from the network adjustment are, in some cases, in very good agreement with the AM0-2 model. The SLR model gives an extremely fast spreading rate between Easter Island and Maui, 143±4 mm/yr, as well as an even larger spreading between Easter Island and Huahine of 172±7 mm/yr. The SL7.1 geodesic rates between Easter Island and Huahine and Maui is, at the 95% confidence level, insignificantly different from that predicted by AM0-2. Although not shown here, the motion components recovered for Huahine appears to have a larger westward component from that expected it were located on a rigid Pacific plate [Smith et al., 1990a]. There has been some recent evidence that the spreading of the plates along at least a portion of the East Pacific Rise is asymmetrical [Hey et al., 1985] and it is possible that our results may reflect a portion of this asymmetry.

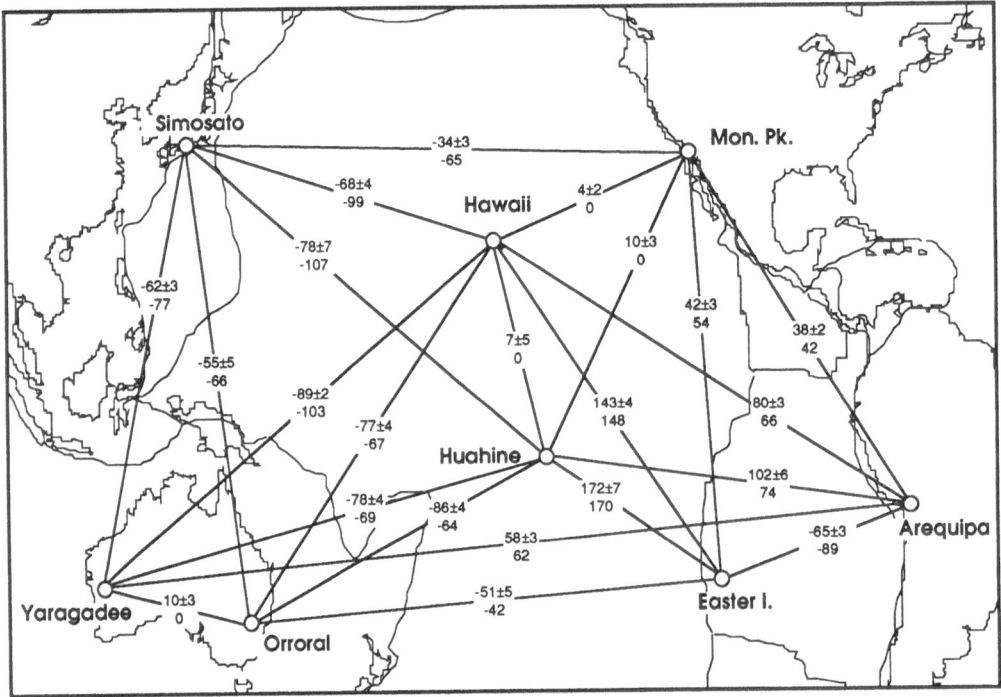

Fig. 8. Geodesic rates between tracking sites across the Pacific Basin in mm/yr.
Top value is determined from the SLR analysis. Bottom value is that predicted by
Minster & Jordan AM0-2 model.

In North America, there are six SLR sites with an observing history sufficient for
inclusion in the present discussion. These sites and some of the geodesic rates between
them are shown in figure 9, along with the AM0-2 rates for interplate lines. Again,
uncertainties assigned to the SLR modeled rates are at the 67% confidence level. The
SLR modeled rate between McDonald and Greenbelt recovered from our model is
8±2 mm/yr, which is anomalously large. Considering only the data obtained since
1985, when the laser at McDonald was upgraded, the observed geodesic rate to
Greenbelt becomes essentially zero. Between Platteville and Greenbelt the rate is -6±5
mm/yr and between Platteville and McDonald, the rate is 1±10 mm/yr. The large
errors associated with Platteville are a due to its short data span. The geodesic rate
across the Basin and Range province, between Platteville and Quincy, as determined
from the SLR model is 12±5 mm/yr. Results from a recent analysis by the GSFC
VLBI group (Ma et al., private communication) indicate rates ranging from 3 mm/yr
between Platteville and Hat Creek (~100 km NNW of Quincy) to 6 mm/yr between
Platteville and Quincy. Although there currently is some disagreement between SLR
and VLBI on this line, spreading thought to be ongoing within the Basin and Range is
indicated by both systems.

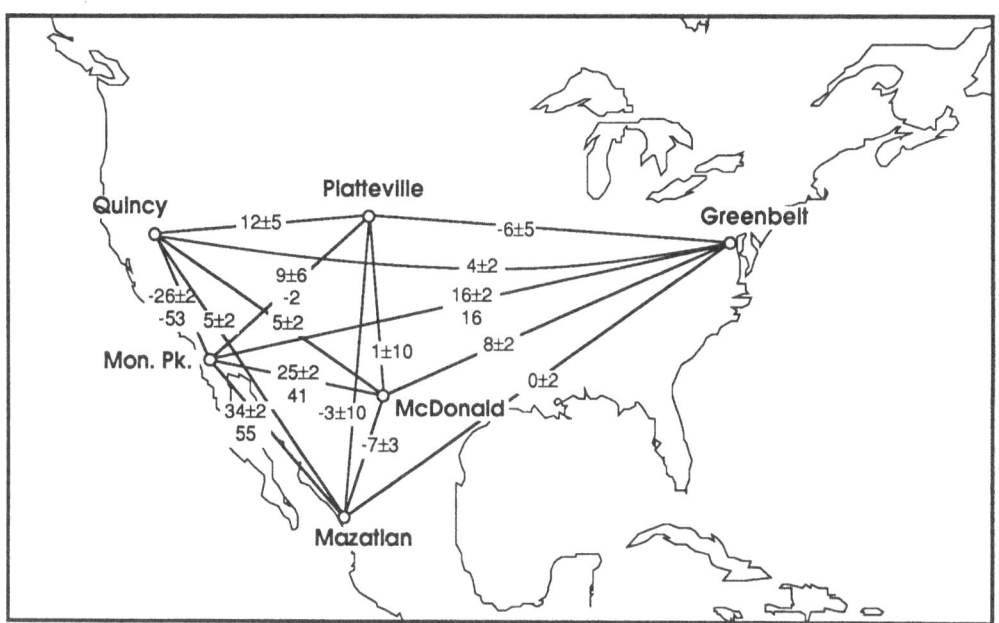

Fig. 9. Geodesic rates between sites in North America in mm/yr. Top value is determined from the SLR analysis. Bottom value is that predicted by Minster & Jordan AM0-2 model. For lines between sites located on the North American plate (where the AM0-2 model predicts zero motion), the bottom value is omitted.

Figure 10 shows modeled SLR and AM0-2 geodesic rates across the North Atlantic for three sites on each plate. Analysis of all reliable Trans-Atlantic SLR rates suggests spreading is a few mm/yr slower than predicted by any the geologic model. The VLBI rates determined by the GSFC analysis group for similar lines also suggest slower relative motion across the Atlantic. If the North American and Eurasian plates are behaving rigidly, then it appears that the results from space geodetic techniques have detected a relatively recent (within the last three million years) change in spreading rate. This question could be better resolved if space geodetic measurements could be taken from sites located immediately east and west of the mid-Atlantic Ridge.

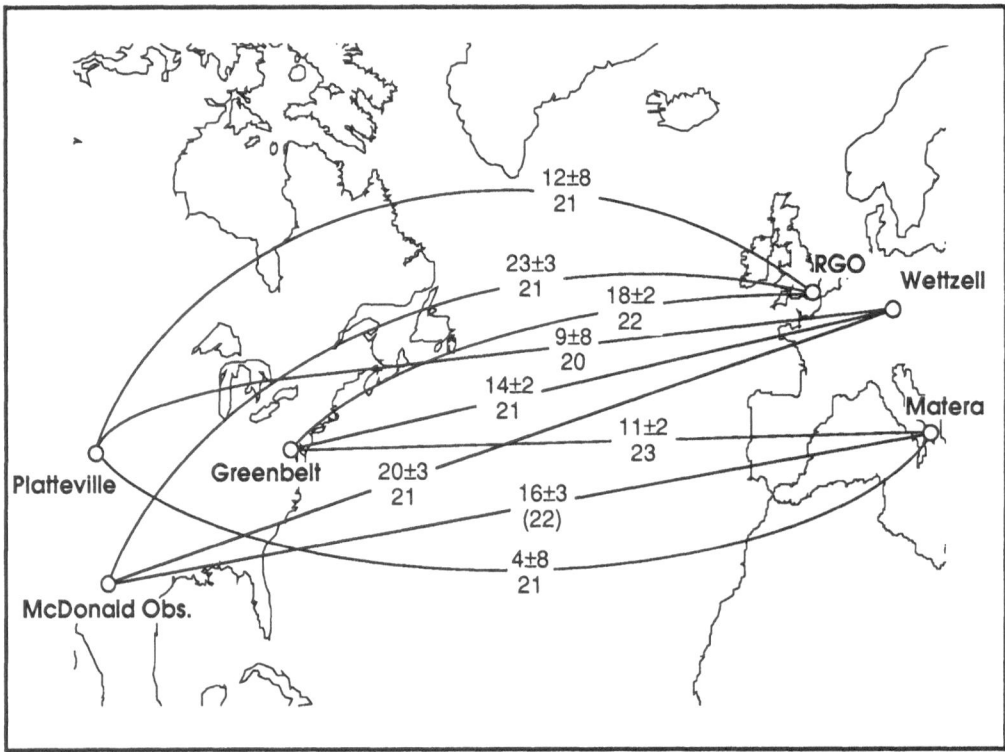

Fig. 10. Geodesic rates across the north Atlantic in mm/yr. Top value is determined from the SLR analysis. Bottom value is that predicted by the Minster & Jordan AM0-2 model.

Figure 11 illustrates the SLR modeled geodesic rates between the northern European sites as well as to Simosato and Arequipa. Intraplate geologic rates are zero by definition and have been omitted.The SLR rates between the northern European sites are essentially zero, with associated uncertainties in the 2-4 mm/yr range (at the 67% confidence level). Potsdam and Grasse suffer largely from an insufficient history of good quality tracking data. We feel that the measures of performance indicated by RGO, Wettzell, and Graz are the more reliable indicators. Independent observations acquired by VLBI between Onsala, Sweden, and Wettzell further support the conclusion of no significant interstation motion.

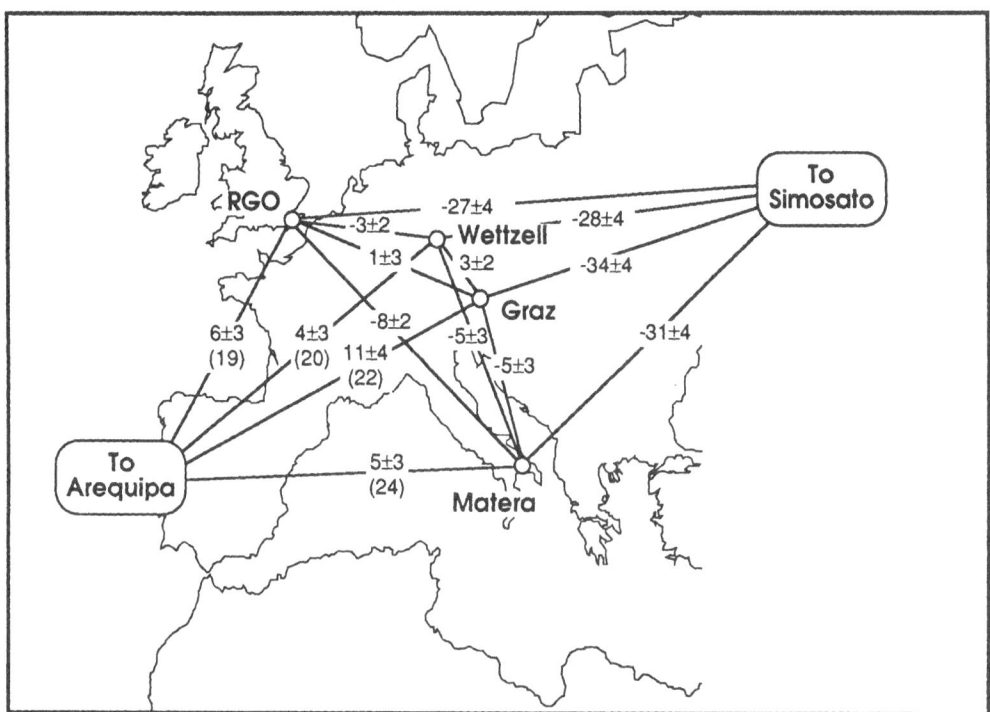

Fig. 11. Geodesic rates for lines between sites in Europe, South America and Japan in mm/yr. Top value is determined from SLR analysis. Bottom value is that predicted by the Minster & Jordan AM0-2 model. For lines between sites located on the Eurasian Plate (where the AM0-2 model predicts zero motion), the bottom value is omitted.

Plate motion models such as AM0-2, NUVEL-1, and Chase [1978] suggest that the relative motion between the African and Eurasian plates consists primarily of north-south convergence. The major seismic activity in western Europe is well known to be located near to and within the Mediterranean Basin, and is largely a consequence of this collision of the African and Eurasian plates. Given that our sites are reasonably representative of the general character of European motion, the observation that no significant intersite motion is detected in northern Europe beyond the 3 mm/yr level, which suggests that this part of Europe is moving as a major block. Simasato has been nominally placed on the Eurasian plate, but it appears to be highly influenced by the tectonics of the Eurasian-Pacific/Philippine plate convergence, and is thus inappropriate to consider for assessing intraplate stability.

SUMMARY

Generally, for sites located within the stable interiors of the major plates, the SLR derived site motions show good agreement with the geological predictions. Mid-Atlantic spreading is observed to be a few mm/yr less than predicted by the rigid-plate motion models. The motion of Yaragadee is about 10 mm/yr less than predicted by any of the geologic models. Rapid spreading across the East Pacific Rise is observed to be about 170 mm/yr. Intraplate spreading within the Basin and Range Province of the United States which is detected at 5 to 10 mm/yr. Strike-slip motion across the San Andreas Fault system is about 20 mm/yr less than predicted by the plate motion models, additional motion may be taking place offshore. Northern Europe appears to be stable, however Matera (Southern Italy) appears to be affected by the motion of the African plate.

REFERENCES

Chase, C. G., The N plate problem of plate tectonics, *Geophys. J. R. Soc., 29*, 117-122, 1972.

Chase, C. G., Plate kinematics: The Americas, East Africa, and the rest of the world, *Earth Planet. Sci. Lett., 37*, 355-368, 1978.

Christodoulidis, D. C., D. E. Smith, R. Kolenkiewicz, P. J. Dunn, S. M. Klosko, M. H. Torrence and S. Blackwell, Observing Tectonic Plate Motions and Deformations from Satellite Laser Ranging, *J. Geophys. Res., 90*, 9249-9263, 1985.

Christodoulidis, D. C., D. E. Smith, R. G. Williamson, and S. M. Klosko, Observed tidal braking in the Earth/Moon/Sun system, *J. Geophys. Res., 93*, 6216-6236, 1988.

Clark, T. A., D. Gordon, W. E. Himwich, C. Ma, A. Mallama, and J. W. Ryan, Determination of relative site motions in the western United States using Mark III Very Long Baseline Interferometry, *J. Geophys. Res. 92*, 12741-12750, 1987.

Cohen, S. C., and D. E. Smith, LAGEOS scientific results: Introduction, *J. Geophys. Res., 90*, 9217-9220, 1985.

DeMets, C., R. G. Gordon, D. F. Argus, and S. Stein, Current plate motions, *Geophys. J. of the RAS, DGG and EGS,* in press, 1989.

Hey, R. N., D. F. Naar, M. C. Kleinrock, W. J. Phipps Morgan, E. Morales, and J.-G. Schilling, Microplate tectonics along a superfast seafloor spreading system near Easter Island, *Nature, 317*, 320-331, 1985.

Jordan, T. H., and J. B. Minster, Measuring crustal deformation in the American west, *Sci. Amer., 259*(8), 48-58, 1988.

Kroger, P. M., G. A. Lyzenga, K. S. Wallace and J. M. Davidson, Tectonic Motion in the Western United States Inferred from Very Long Baseline Interferometry Measurements, 1980-1986, *J. Geophys. Res., 92*, 14151-14163, 1987.

Ma, C., J. W. Ryan, and D. Caprette, Crustal Dynamics Project data analysis - 1988, VLBI geodetic results 1979-87, *NASA Tech. Memo., 100723*, 1989.

Marsh, J. G., F. J. Lerch, B. H. Putney, D. C. Christodoulidis, D. E. Smith, T. L. Felsentreger, B. V. Sanchez, S. M. Klosko, E. C. Pavlis, T. V. Martin, J. W. Robbins, R. G. Williamson, O. L. Colombo, D. D. Rowlands, W. F. Eddy, N. L. Chandler, K. E. Rachlin, G. B. Patel, S. Bhati and D. S. Chinn, A new gravitational model for the Earth from satellite tracking data: GEM-T1, *J. Geophys. Res., 93*, 6169-6215, 1988.

Martin, T. V., W. F. Eddy, D. D. Rowlands, and D. E. Pavlis, GEODYN-II system operations manual, vol. 1-5, contractor report, EG&G-Washington Anal. Serv. Center, Lanham, MD, April 1987.

Melbourne, W., R. Anderle, M. Feissel, R. King, D. McCarthy, D. Smith, B. Tapley, and R. Vicente, Project MERIT standards, Circ. *167*, U. S. Naval Obs., Washington, D. C., 1983.

Minster, J. B., and T. H. Jordan, Numerical modelling of instantaneous plate tectonics, *Geophys. J. R. astr. Soc., 36*, 541-576, 1974.

Minster, J. B., and T. H. Jordan, Present-day plate motions, *J. Geophys. Res., 83*, 5331-5254, 1978.

Minster, J. B., and T. H. Jordan, Vector constraints on western U. S. deformation from space geodesy, neotectonics, and plate motions, *J. Geophys. Res., 92*, 4798-4804, 1987.

Putney, B. H., General theory for dynamic satellite geodesy, The National Geodetic Satellite Program, *NASA Spec. Pub., SP-365*, 319-334, 1977.

Schutz, B. E., M. M. Watkins, R. J. Eanes, and B. D. Tapley, Global plate motions derived from LAGEOS laser ranging, (abstract), *EOS, 70*, 304, 1989.

Smith, D. E., R. Kolenkiewicz, M. H. Torrence, P. J. Dunn, R. G. Williamson, S. M. Klosko, J. W. Robbins, E. C. Pavlis, and S. K. Fricke, Mediterranean geodynamics from satellite laser ranging, (abstract), *EOS, 70*, 305, 1989a.

Smith, D. E., R. Kolenkiewicz, B. H. Putney, P. J. Dunn, S. M. Klosko, E. C. Pavlis, J. W. Robbins, M. H. Torrence, R. G. Williamson, and S. K. Fricke, A geodetic Earth motion model derived from LAGEOS observations: GSFC-SL7, *NASA Tech. Memo.*, in preparation, 1989b.

Torrence, M. H., S. M. Klosko, and D. C. Christodoulidis, The construction and testing of normal points at Goddard Space Flight Center, in *Proceedings of the Fifth international workshop on laser ranging instrumentation*, ed. by J. Gaignebet, 506-516, Geodetic Institute, Bonn, 1984.

Nonlinear Inversion of Geodetic and Geophysical Data : Diagnosing Nonlinearity

P.J.G. Teunissen
Geodetic Centre
Delft University of Technology
The Netherlands

ABSTRACT:

This paper addresses the problem of diagnosing nonlinearity in the nonlinear inversion of geodetic and geophysical data. Measures of nonlinearity are proposed that can be used to assess the amount of nonlinearity in nonlinear models and to test whether a linear (ized) model is a sufficient approximation.

After the introductory section, which gives a brief overview of the various problems associated with nonlinear inversion, section two discusses three simple measures of nonlinearity that can be used as a first step in analyzing the amount of nonlinearity of the model. In section three we show that in the problem of nonlinear inversion one has to reckon with two different types of nonlinearity. First of all there is the nonlinearity of the parameter-curves which obviously depends on the chosen parametrization. Secondly there is the nonlinearity related to the curvature of the manifold. It is intrinsic in the sense that it is independent of the choice of parametrization. The two types of nonlinearity are described using concepts from differential geometry.

In section four we discuss the geometry of nonlinear least-squares inversion. In this section it is also shown how the above mentioned two types of nonlinearity affect the first moments of the nonlinear least-squares estimators. Finally in section five a strategy is proposed for diagnosing the significance of nonlinearity in nonlinear models.

1 INTRODUCTION

Nonlinear optimization, nonlinear least-squares and densities of nonlinear estimators are a trilogy of problems that are intimately related in the framework of *nonlinar inversion* of geodetic and geophysical data [Teunissen, 1985; Tarantola, 1987]. Usually the description of physical phenomena proceeds through models in which a mapping, A, is defined, from a set of parameters, N, to a set of experimental outcomes, M. M is supposed to contain the image of the map A. We will assume that $M = R^m$ and $N = R^n$. Obtaining the image $y = A(x) \in M$ of $x \in N$ is solving the *forward problem*. Obtaining the parameters $x \in N$ that correspond to $y \in M$ in e.g. a least-squares sense, is solving the *inverse problem*. The inverse problem is said to be linear if $A(\alpha_1 x_1 + \alpha_2 x_2) = \alpha_1 A(x_1) + \alpha_2 A(x_2), \forall \alpha_1, \alpha_2 \in R; x_1, x_2 \in N$. Almost no geodetic or geophysical inverse problem is truely linear. A consequence of nonlinearity is that the inverse problem increases in complexity. These complications manifest themselves: a) in the problem of finding the numerical estimates of the parameters x; and b) in the problem of finding the a posteriori probability density function of the nonlinear estimators.

The numerical estimation of parameters is typically a problem of optimization. The estimation of parameters requires frequently the maximization or minimization of an objective function. Typical objective functions are risk functions, robust loss functions, posteriori density functions, likelihood functions and (weighted or unweighted) sums of squares. In general no direct methods exist for estimation in nonlinear models. For these cases the nonlinear problem is attacked iteratively: at each step the solution af a linear problem, in terms of (the Fréchet) derivatives, is constructed. Various iterative techniques exist for solving nonlinear optimization problems. The best known iterative techniques are: the Steepest-Ascent (Descent) method [Cauchy, 1847], the (Quasi-) Newton methods [Fletcher and Powell, 1963; Broyden, 1967], the Conjugate Direction methods [Fletcher and Reeves, 1964], and the Trust-Region methods [Levenberg, 1944; Marquardt, 1963; Goldfeld et al., 1966]. For a survey of these techniques see e.g. [Ortega and Rheinboldt, 1970] and [Bard,1974].

Numerical studies of nonlinear geodetic inverse problems solved using the least-squares criterion can be found in [Kubik, 1967; Pope, 1972; Saito, 1973; Stark and Mikhail, 1973; Pope 1974; Schek and Maier, 1976; Kelley and Thompson, 1978; Teunissen, 1984; Blaha, 1987]. Similar studies for the numerical nonlinear inversion of geophysical data can be found in [Jupp and Vozoff, 1975; Tarantola and Valette, 1982; Oldenburg, 1983; Tarantola, 1986; Tarantola et al., 1987; Williams and Richardson, 1988]. For problems of reentry space vehicle tracking and model deformation studies based on nonlinear time series analysis or state-space dynamic models the reader is referred to [Denham and Pines, 1966; Larson et al., 1967; Mehra, 1971; Liang and Christensen, 1975; Sorenson, 1977; Krebs, 1980; Austin and Leondes, 1981; Haggan and Ozaki, 1981; Keenan, 1985; Tong, 1986]. Although most nonlinear inverse problems have to be attacked iteratively, there exist particular cases which can be solved in closed form, see e.g. [Sanso, 1973; Teunissen, 1987, 1988; Grafarend et al., 1989; Grafarend and Schaffrin, 1989; Koch, 1989].

It will be clear that a numerical parameter estimation or inversion procedure is incomplete without an analysis of the uncertainties in the results. That is, it is not enough to compute the nonlinear parameter estimates and state that they are the solution to the inverse problem. Knowledge of the a posteriori probability density functions of the nonlinear estimators is needed in order to infer the quality of the results obtained. For linear models a rather complete theory of inference exists [see e.g. Baarda, 1967; Mood et al., 1974; Graybill, 1976; Koch, 1988]. Unfortunately the results which hold true for linear models do not carry over to the nonlinear case. That is, although some exact methods for deriving the distribution of nonlinear estimators exist, these methods are in general very difficult to apply in practice.

A first approach that comes to mind to compute the distribution of nonlinear estimators is based on the fundamental relations that define distribution functions. If the parent density is given, then theoretically at least, one can find both the cumulative and density distribution of the nonlinear estimator. The practical problem with this method is however that in general one cannot easily evaluate the complicated integrals and inverses of the nonlinear maps involved. Instead of aiming at a complete description of the distribution, one could restrict oneselves to some of the moments of the distribution. The complexity of these computations depends very much on the nature of the parent distribution and the nonlinear maps involved. But in general they can become quite complicated, especially in the multivariate case.

If in a particular problem it is impossible to apply the above mentioned analytical methods, the next one thing one can try to do is to make use of approximations based on a suitable Taylor expansion. In this way appropriate approximations to the first two moments and density of nonlinear least-squares estimators were obtained in [Wolf, 1961; Schaffrin, 1983; Teunissen, 1984, Pazman, 1987; Jeudy, 1988; Teunissen 1988a, 1988b]. In this context geometrical tools from differential geometry have proven to be very useful [Krarup, 1982; Teunissen, 1985; Borre and Lauritzen, 1987]. A review of the use of differential geometry in statistical inference can be found in [Amari et al., 1987]. An analytical expression for the first moment of the nonlinear least-squares estimators of the parameters of the Symmetric Helmert Transformation is given in [Teunissen, 1989].

An alternative way to estimate the distribution of nonlinear estimators would be to rely on Monte Carlo methods [e.g. Koch, 1989]. One replicates the series of experiments as many times as one needs, each time with a new sample drawn from the parent distribution and so obtains the relevant distributional properties by averaging over all replications. A possible drawback of this technique is however that it may become computationally demanding for large scale inverse problems.

Finally another way to estimate the properties of nonlinear estimators is to rely on the results from asymptotic theory [Jennrich, 1969; Schmidt, 1982; Bierens, 1984]. The central idea of asymptotic theory is that when the number of observations is large and errors of estimation corresponding small, simplifications become available that are not available in general. The rigorous mathematical development involves limiting distributional results and is closely related to the classical limit theorems of probability theory. Unfortunately, since the theory is based on the assumption that the number of observations increase indefinitely, the results obtained up to now cannot satisfy all the requirements of application in practice.

From the above given general remarks it will be clear that the solution of the nonlinear inverse problem is not as straightforward as it is for the linear case. It is therefore expedient to have ways of assessing the amount of nonlinearity in nonlinear models and methods to prove whether a linear(ized) model is a sufficient approximation. A first attempt to examine the influence of nonlinearity in the dependence of data on parameters for geophysical models was made by [Kennett, 1978]. Similar studies for geodetic models can be found in [Teunissen, 1985a; Bähr, 1985, 1988]. The influence of nonlinearity on the solution of the geodetic boundary value problem was investigated in [Heck, 1988].

In the present contribution we will study ways of diagnosing nonlinearity. In section 2 we start with three simple measures of nonlinearity which can be used to analyze the amount of nonlinearity of the model. In section 3 we will show that one has to discriminate between two types of nonlinearity, an intrinsic nonlinearity and a nonlinearity that depends on the chosen parametrization. These two types of nonlinearity are described using concepts from differential geometry. Our geometric approach is continued in section 4 where the geometry of nonlinear least-squares inversion is discussed. As a result two simple expressions for the nonlinearity of the manifold and the nonlinearity in the parameter curves are derived. Finally in section 5 we show how these two measures can be used for testing the significance of nonlinearity in the problem of nonlinear least-squares inversion.

2 THREE SIMPLE MEASURES OF NONLINEARITY

We consider the nonlinear model

$$E\{\underline{y}\} = A(x) \quad ; \quad D\{\underline{y}\} = \sigma^2 Q_y , \tag{1}$$

where $E\{.\}$ and $D\{.\}$ are the operators of mathematical expectation and dispersion respectively; $\underline{y}$ is the random data vector of size m; $A(.)$ is a nonlinear map from R^n into R^m; x is the unknown parameter vector of size n; σ^2 is the variance factor of unit weight; and $\sigma^2 Q_y$ is the positive definite variance-covariance matrix of $\underline{y}$.

Since, in most applications one to a large extend relies on the results from the theory of *linear* inference we oppose the nonlinear model (1) to its linearized version

$$E\{\Delta \underline{y}\} = \partial_x A(x_0)\Delta x \quad ; \quad D\{\Delta \underline{y}\} = \sigma^2 Q_y , \tag{2}$$

where $\Delta \underline{y} = \underline{y} - A(x_0); \Delta x = x - x_0$ and $\partial_x A(x_0)$ is the matrix of partial derivatives of $A(.)$ evaluated at $x_0 \in R^n$. The approximations involved in replacing the nonlinear model (1) by the linear model (2) follow from *Taylor's formula with remainder*. Let $F : R^n \to R$ be a function of class C^q. Taylor's formula with remainder reads then

[Loomis and Sternberg, 1968]:

$$F(x) = F(x_0) + \partial_{\alpha_1} F(x_0) \Delta x^{\alpha_1} + \frac{1}{2!} \partial^2_{\alpha_1 \alpha_2} F(x_0) \Delta x^{\alpha_1} \Delta x^{\alpha_2} + \cdots$$
$$\cdots \frac{1}{(q-1)!} \partial^{q-1}_{\alpha_1 \cdots \alpha_{q-1}} F(x_0) \Delta x^{\alpha_1} \cdots \Delta x^{\alpha_{q-1}} + R_q(x) , \tag{3}$$

where Einstein's summation convention is used, $\alpha_1, \alpha_2, \ldots = 1, \ldots, n$, $\Delta x = x - x_0$ and

$$R_q(x) = \frac{1}{q!} \partial^q_{\alpha_1 \cdots \alpha_q} F(x_0 + t\Delta x) \Delta x^{\alpha_1} \cdots \Delta x^{\alpha_q} , \tag{4}$$

with $t \in (0, 1)$.

Application of Taylor's formula with remainder to (1) and (2) shows that the second order remainder

$$R_2(x) = \frac{1}{2} \Delta x^* \partial^2_{xx} A(x_0 + t\Delta x) \Delta x , \tag{5}$$

is neglected in (2). A bound on this remainder may therefore be used as a first measure of nonlinearity. We will first give an explicit estimate of a bound on (4), and then specialize to the case of (5).

Suppose that all q-th order partial derivatives of F satisfy

$$| \partial^q_{\alpha_1 \cdots \alpha_q} F(x) | \le c \in R . \tag{6}$$

Then with (4),

$$| R_q(x) | \le \frac{c}{q!} \sum_{\alpha_1 \cdots \alpha_q = 1}^{n} | \Delta x^{\alpha_1} | \cdots | \Delta x^{\alpha_q} | . \tag{7}$$

Since,

$$\sum_{\alpha_1 \cdots \alpha_q = 1}^{n} \Delta x^{\alpha_1} \cdots \Delta x^{\alpha_q} = [\Delta x^1 + \Delta x^2 + \cdots + \Delta x^n]^q , \tag{8}$$

for any real numbers $\Delta x^1, \ldots, \Delta x^n$, (8) can be written as

$$| R_q(x) | \le \frac{c}{q!} \left[\sum_{\alpha=1}^{n} | \Delta x^{\alpha} | \right]^q . \tag{9}$$

And since,

$$\sum_{\alpha=1}^{n} | \Delta x^{\alpha} | \le n^{1/2} \| \Delta x \| ,$$

it follows from (9) that the remainder in Taylor's formula satisfies the estimate

$$| R_q(x) | \le \frac{c}{q!} n^{q/2} \| \Delta x \|^q . \tag{10}$$

Application of this result to (5) gives the following bound on the second order remainder $R_2(x)$:

$$\mid R_2^i(x) \mid \leq \frac{c^i}{2} n \parallel \Delta x \parallel^2 \quad , \quad if \quad \mid \partial_{\alpha\beta}^2 A^i(x) \mid \leq c^i \quad , \qquad \begin{matrix} i = 1, \ldots, m \\ \alpha, \beta = 1, \ldots, n \end{matrix} \tag{11}$$

Since the evaluation of the scalars $c^i, i = 1, \ldots, m$ is based on the individual elements of the Hessian matrices of the observation equations, the computation of the upperbound (11) is in general not too difficult. A disadvantage of the above upperbound is that it can become somewhat pessimistic for large n, e.g. in case of many parameters.

An alternative bound for the remainder $R_2(x)$ may be based on the extreme eigenvalues of the Hessian matrix $\partial_{xx}^2 A^i(x)$. This bound follows from the *Rayleigh Quotient Theorem* [Ortega, 1987]:

Let Q be a real $n \times n$ symmetric matrix with eigenvalues $\lambda_1 \leq \cdots \leq \lambda_n$. Then the minimum and maximum values of the Rayleigh quotient $x^* Q x / x^* x$ are λ_1 and λ_n, respectively, and these values are taken on for any eigenvector of Q corresponding to λ_1 or λ_n, respectively.

Application of this theorem to the remainder $R_2(x)$ gives the interval

$$\frac{1}{2} \lambda_1^i \parallel \Delta x \parallel^2 \leq R_2^i(x) \leq \frac{1}{2} \lambda_n^i \parallel \Delta x \parallel^2 \quad , \quad i = 1, \ldots, m \tag{12}$$

Note that these bounds are the sharpest ones available.

In general the computation of eigenvalues can become quite tedious. The computation of the extreme eigenvalues simplifies however if the matrix $\partial_{xx}^2 A^i(x)$ is sparce, i.e. if only a few parameters are involved per observation equation. This is usually the case in geodetic applications. Take for instance the distance-observation equation $l_{ij} = (x_{ij}^2 + y_{ij}^2)^{1/2}$ where l_{ij} is the distance between two points i and j, and x_{ij} and y_{ij} are the corresponding cartesian coordinate differences. The Hessian matrix reads then

$$\partial_{\alpha\beta}^2 l_{ij} = \frac{1}{l_{ij}^3} \begin{bmatrix} y_{ij}^2 & -x_{ij} y_{ij} & -y_{ij}^2 & x_{ij} y_{ij} \\ -x_{ij} y_{ij} & x_{ij}^2 & x_{ij} y_{ij} & -x_{ij}^2 \\ -y_{ij}^2 & x_{ij} y_{ij} & y_{ij}^2 & -x_{ij} y_{ij} \\ x_{ij} y_{ij} & -x_{ij}^2 & -x_{ij} y_{ij} & x_{ij}^2 \end{bmatrix}$$

The extreme eigenvalues of this matrix are easy to compute. They are

$$\lambda_1 = 0 \quad and \quad \lambda_n = 1/l_{ij}$$

Thus, for the second order remainder of the distance equation the following bounds hold:

$$0 \leq R_2 \leq (\Delta x_{ij}^2 + \Delta y_{ij}^2)/2l_{ij} \tag{13}$$

Some numerical values of this interval are given in table 1.

$\Delta x_{ij}, \Delta y_{ij}$	l_{ij}	R_2
100 m	1 km	$\leq$ 20. m
50 m	1 km	$\leq$ 2.5 m
10 m	1 km	$\leq$ 0.1 m
5 m	1 km	$\leq$ .03 m

Table 1: Bounds on the second order remainder R_2 of $l_{ij} = (x_{ij}^2 + y_{ij}^2)^{1/2}$.

A third way to measure the nonlinearity of the observation equations would be to take an (unweighted or weighted) average of the remainder $R_2(x)$. For this approach we will make use of the following theorem [Koch, 1988]:

Let $\underline{x}$ be a random vector of size n, with mean $E\{\underline{x}\} = x$ and dispersion $D\{\underline{x}\} = Q_x$. Then

$$E\{\underline{x}^* Q \underline{x}\} = trace[QQ_x] + x^* Q x , \tag{14}$$

where Q denotes the symmetric $n \times n$ matrix of the quadratic form .

In order to apply this theorem to the second order remainder $R_2(x)$, we will *assume* that $\Delta \underline{x}$ is a random n-vector with zero mean and covariancematrix $Q_{\Delta x}$. The mean of the remainder $\underline{R}_2(x)$ follows then as

$$E\{\underline{R}_2^i(x)\} = \frac{1}{2} trace[\partial_{xx}^2 A^i(x) Q_{\Delta x}], \quad i = 1, \ldots, m \tag{15}$$

Note that this average measure of nonlinearity is very easy to compute if we take a scaled identity matrix, $\sigma^2 I_n$, for $Q_{\Delta x}$. For the distance-observation equation this would give

$$E\{\underline{R}_2\} = \sigma^2/l_{ij} .$$

Also note that (15) is independent of the distribution assumed for $\underline{\Delta x}$.

The above measure of nonlinearity differs in two ways from the previous two measures, (11) and (12). First of all the measure (15) is a probabilistic measure in contrast to the deterministic measures (11) and (12). The measure (15) allows one to estimate the bias, due to nonlinearity, in $\underline{y}$ if (1) would be solved as the forward problem $\underline{y} = A(\underline{x})$. Secondly the measure (15) differs from (11) and (12) in that it is an average measure. The measure (15) may therefore, for a particular case, underestimate the nonlinearity present. This should however not be seen as a serious drawback, because of the random nature of $\underline{y}$. In the following sections we will see that a formula similar to (15) can be used as an upperbound on the biases of least-squares estimators.

The above given three measures of nonlinearity can be used as a first step in analyzing the amount of nonlinearity of the model (1). It should be noted however that all three measures of nonlinearity depend on the chosen parametrization. That is, they are *not* invariant under a change of variables in $A(x)$. To some extent this is also as it should be. Reparametrization, i.e. replacing the parameters x by new parameters $\bar{x}$, implies namely the introduction of a new one-to-one map $x(\bar{x})$ from R^n to R^n with possibly additional nonlinearity. However, we also know that a reparametrization does *not* affect the solution of the nonlinear inverse problem in the *dataspace* R^m. It seems therefore that we have to discriminate between two types of nonlinearity. How this can be done is made clear in the next section.

3 TWO TYPES OF NONLINEARITY

We rewrite the nonlinear model (1) as

$$\underline{y} = A(x) + \underline{e} \quad ; \quad D\{\underline{y}\} = \sigma^2 Q_y , \tag{16}$$

where $E\{\underline{e}\} = 0$.

For varying values of x, $A(x)$ traces locally an n-dimensional surface or manifold embedded in R^m. The functional equation of (16) can therefore be described geometrically as shown in figure 1.

From figure 1 we learn that we have to discriminate between two types of nonlinearity. First of all there is the nonlinearity of the parameter curves or coordinate lines in the manifold. This type of nonlinearity obviously depends on the way the manifold is parametrized. The second type of nonlinearity is related to the nonlinearity of the manifold itself. It is intrinsic in the sense that it is independent of the choice of parametrization. Just like a circle remains a circle, whether it is parametrized with cartesian- or with polar coordinates.

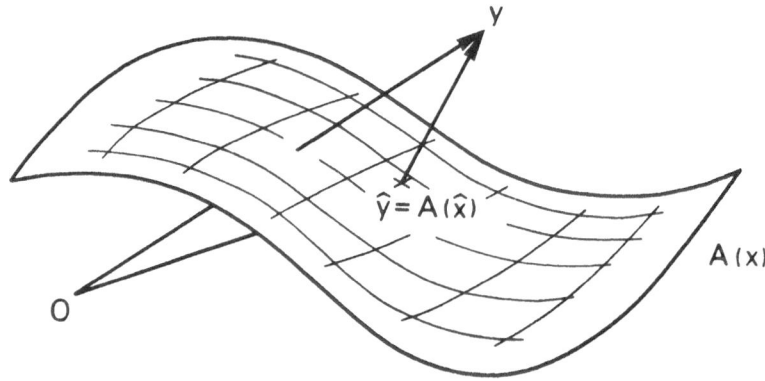

Figure 1: The geometry of $\underline{y} = A(x) + \underline{e}$.

With the given geometric interpretation it seems natural to introduce concepts of differential geometry for describing the above mentioned two types of nonlinearity. We will first consider the nonlinearity of the manifold. The nonlinearity of the manifold can be defined as the deviation of the manifold from its tangentspace in the neighborhood of the point of tangency. It is convenient to introduce the function

$$d(x) = n^* Q_y^{-1}[A(x) - A(x_0)] , \tag{17}$$

where n is a unitvector normal to the tangentspace of the manifold at the point $A(x_0)$. Thus

$$n^* Q_y^{-1} \partial_x A(x_0) = 0 \quad and \quad n^* Q_y^{-1} n = 1 . \tag{18}$$

Note that we assume the metric of R^m to be described by the positive-definite matrix Q_y^{-1}.

The function $d(x)$ describes the perpendicular distance along n from the tangentspace to the point $A(x)$ on the manifold. If we assume as before that the map $A(x)$ is sufficiently smooth, substitution of its Taylor expansion at x_0 in (17) gives with (18),

$$d(x) = \frac{1}{2}\Delta x^* [n^* Q_y^{-1} \partial_{xx}^2 A(x_0)]\Delta x + \cdots \tag{19}$$

The *second fundamenal form* of the manifold is defined as the quadratic form

$$II = v^*[n^*Q_y^{-1}\partial_{xx}^2 A(x)]v \quad , \quad v \in R^n \quad , \quad n \in R(\partial_x A(x))^\perp \subset R^m \tag{20}$$

Thus for small values of v the function $2d(x)$ can be approximated by the second fundamental form II with errors of third or higher orders in v. A study of the form II will therefore give information about the shape of the manifold $A(x)$ near the point of tangency.

With the second fundamental form it is now a small step to introduce the concept of *normal curvature*. In Gaussian surface theory the normal curvature is defined as the ratio of the second fundamental form and the first fundamental form [Spivak, 1979]. The *first fundamental form* is defined as

$$I = v^*Q(x)^{-1}v \quad , \quad with \quad Q(x)^{-1} = \partial_x A(x)^* Q_y^{-1} \partial_x A(x) . \tag{21}$$

With the first fundamental form one can compute the arclength of a curve in the manifold. The matrix $Q(x)^{-1}$ is known as the induced metric of the manifold. Note that $Q(x)^{-1}$ corresponds to the *normal matrix* of the problem of linearized least-squares inversion.

With (20) and (21) the normal curvature becomes

$$k_n(v) = \frac{II}{I} = \frac{v^*[n^*Q_y^{-1}\partial_{xx}^2 A(x)]v}{v^*Q(x)^{-1}v} \tag{22}$$

The normal curvature is invariant under a change of variables in $A(x)$. This can be seen as follows. Let $x(\bar{x})$ be a one-to-one map from R^n to R^n. Then

$$\begin{aligned}
\partial_{\bar{x}\bar{x}}^2 A(\bar{x}) &= \partial_{\bar{x}}x^* \partial_{xx}^2 A(x(\bar{x}))\partial_{\bar{x}}x + \partial_x A(x(\bar{x}))\partial_{\bar{x}\bar{x}}^2 x \\
Q(\bar{x})^{-1} &= \partial_{\bar{x}}x^* Q(x(\bar{x}))^{-1}\partial_{\bar{x}}x \\
v &= \partial_{\bar{x}}x\bar{v} .
\end{aligned}$$

Substitution into

$$k_n(\bar{v}) = \frac{\bar{v}^*[n^*Q_y^{-1}\partial_{\bar{x}\bar{x}}^2 A(\bar{x})]\bar{v}}{\bar{v}^*Q(\bar{x})^{-1}\bar{v}}$$

shows then, since $n^*Q_y^{-1}\partial_{\bar{x}}A(x(\bar{x})) = 0$, that $k_n(\bar{v}) = k_n(v)$. This invariance of the normal curvature under a change of variables implies that the curvature $k_n(v)$ is an intrinsic property of the manifold $A(x)$ embedded in R^m.

Formula (22) can be used in different ways for measuring the nonlinearity of the manifold. The curvature of the manifold corresponding to a particular direction v in the tangentspace and a particular normal direction n in the orthogonal complement of the tangentspace is given by (22). If one wants the curvature of the manifold for a particular parameter curve, say the αth-coordinate line, then (22) simplifies to

$$k_n(\alpha) = \frac{[n^* Q_y^{-1} \partial_{xx}^2 A(x)]_{\alpha\alpha}}{N_{\alpha\alpha}} \,, \tag{23}$$

where $N_{\alpha\alpha}$ is the αth diagonal element of the normal matrix.

In order to determine the extreme values of the normal curvature, the so-called *principal normal curvatures*, we have to proceed as follows. Since (22) is a generalized Rayleigh quotient the extreme values of the normal curvature follow as the eigenvalues of the generalized eigenvalue problem

$$| n^* Q_y^{-1} \partial_{xx}^2 A(x) - \lambda Q(x)^{-1} | = 0 \tag{24}$$

Note that since $dimR(\partial_x A(x)) = n$ and $dimR(\partial_x A(x))^{\perp} = m - n$, the number of principal normal curvatures equals $n(m - n)$. We will denote the n-number of principal normal curvatures for the normal direction n by

$$k_n^1 \le k_n^2 \le \cdots \le k_n^n \,. \tag{25}$$

In general the computation of the principal normal curvatures can become quite involved. In [Teunissen, 1985] it is shown however that one need not always compute the principal normal curvatures explicitly, since numerical estimates of them become readily available during the Gauss-Newton iteration of the nonlinear least-squares inverse problem.

Instead of using the individual principal normal curvatures one may also use the average principal normal curvatures as measure of nonlinearity of the manifold. The average principal normal curvature for the normal direction n reads

$$\bar{k}_n = \frac{1}{n} \sum_{\alpha=1}^{n} k_n^{\alpha} \,. \tag{26}$$

Since the sum of the n eigenvalues of (24) equals the sum of the n diagonal entries of the matrix $n^* Q_y^{-1} \partial_{xx}^2 A(x) Q(x)$, equation (26) can be written as

$$\bar{k}_n = \frac{1}{n} trace[n^* Q_y^{-1} \partial_{xx}^2 A(x) Q(x)] \,. \tag{27}$$

It will be clear that the computation of this average curvature is much simpler than the computation of the individual principal normal curvatures. When we compare (27) with (15) we note a striking resemblance between the two formulae. In fact if $Q_{\Delta x}$ in (15) is chosen equal to $Q(x)$ the average curvature $\bar{k}_n$ is seen to be equal to two times over n times the normal component of $E\{\underline{R}_2\}$. We will discuss this resemblance further in section 4.

Now that we have discussed the nonlinearity of the manifold we turn our attention to the curvature of the parameter curves. We know that the nonlinearity of the nonlinear map $A(x)$ is essentially described by $\partial^2_{xx} A(x)$. We also know from our discussion above that the intrinsic nonlinearity of the manifold is described by the *normal* component of $\partial^2_{xx} A(x)$. In fact if the manifold is flat then the second fundamental form II vanishes and thus also the normal component of $\partial^2_{xx} A(x)$ vanishes. The remaining nonlinearity due to the nonlinear parameter curves in the flat manifold must then be described by the tangential component of $\partial^2_{xx} A(x)$.

In analogy to (22) the nonlinearity of the parameter curves can be described by

$$\boxed{\gamma_t(v) = \frac{v^*[t^* Q_y^{-1} \partial^2_{xx} A(x)]v}{v^* Q(x)^{-1} v}} \, , \tag{28}$$

where t is a vector lying in the rangespace of $\partial_x A(x)$. It can be shown [Teunissen, 1985] that (28) is closely related to the concept of *covariant differentiation* of differential geometry. Note that (28), in contrast to (22), is not invariant under a change of variables in $A(x)$.

Instead of $\gamma_t(v)$ one may also use the extreme values of the generalized Rayleigh quotient (28) as measures of the nonlinearity of the parameter curves. And in analogy to (27) an average measure of the nonlinearity of the parameter curves is

$$\bar{\gamma}_t = \frac{1}{n} trace[t^* Q_y^{-1} \partial^2_{xx} A(x) Q(x)] \, . \tag{29}$$

The above given measures of nonlinearity can be used to analyze the two existing types of nonlinearity: 1) nonlinearity of the manifold; 2) nonlinearity of the parameter curves. These measures are however, just like the ones of section 2, independent of the chosen method for nonlinear inversion. They do therefore not describe what the influence of nonlinearity is on the result of estimation. It will be clear that this influence may vary from one inversion procedure to another. In most geodetic and geophysical models the inverse problem is solved using the criterion of least-squares [Teunissen, 1985; Tarantola,

1987]. In the next section we will therefore discuss the problem of nonlinear least-squares inversion.

4 GEOMETRY OF NONLINEAR LEAST-SQUARES INVERSION

The problem of nonlinear least-squares can be formulated as the minimization problem:

$$\min_x \parallel y - A(x) \parallel , \tag{30}$$

where $\parallel . \parallel^2 = (.)^* Q_y^{-1}(.)$.

If the metric of the dataspace R^m is described by the positive-definite matrix Q_y^{-1}, the scalar $\parallel y - A(x) \parallel$ equals the distance from the datapoint y to the point $A(x)$ on the manifold. Hence, the problem of (30) corresponds to the problem of finding that point on the manifold, say $\hat{y} = A(\hat{x})$, which has least distance to y. This geometry of the nonlinear least-squares problem is sketched in figure 2.

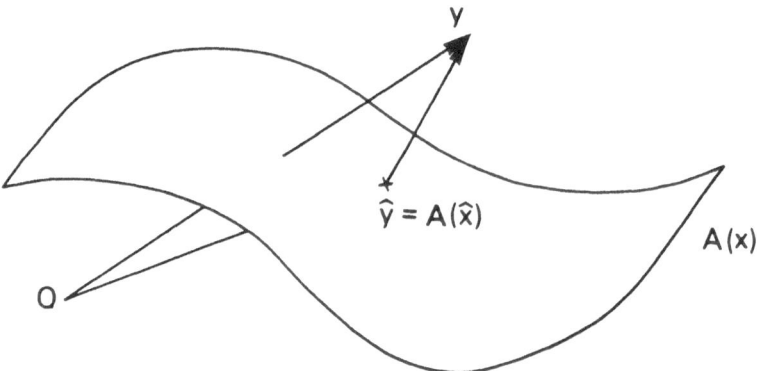

Figure 2: Geometry of nonlinear least-squares.

As shown in [Teunissen, 1985] the necessary and sufficient conditions for $\hat{y} = A(\hat{x})$ to be the solution of the nonlinear least-squares inverse problem, can be formulated in geometric terms as

$$
\begin{array}{ll}
a) & e(\hat{x}) \perp R(\partial_x A(\hat{x})) \\
b) & k_{\hat{n}}^n \parallel e(\hat{x}) \parallel < 1
\end{array}
\tag{31}
$$

where

$$e(\hat{x}) = y - A(\hat{x}) \quad and \quad \hat{n} = \frac{e(\hat{x})}{\parallel e(\hat{x}) \parallel} .$$

Equation (31a) states that the least-squares residual vector should be othogonal to the tangentspace of the manifold at the solution $\hat{y} = A(\hat{x})$. And equation (31b) states that the datapoint y should lie within a hypersphere with centre $\hat{y}$ and a radius equal to the largest principal normal curvature corresponding with the normal direction $\hat{n}$. Note that both conditions of (31) are invariant under a change of variables.

We will use (31) in the development of our measures of nonlinearity for the problem of nonlinear least-squares inversion. We will assume that the nonlinear least-squares estimator $\hat{x}$ can be written as a smooth map of the random datavector y:

$$\hat{x} = F(y), \quad F : R^m \to R^n . \tag{32}$$

We also assume that condition (31b) is satisfied and that $x = F(E\{y\})$. If we Taylorize (32) at the mean of y, we get an expansion in $e = y - E\{y\}$:

$$\hat{x} = x + \partial_y F e + \frac{1}{2} e^* \partial_{yy}^2 F e + \cdots \tag{33}$$

Since $E\{e\} = 0$ and $E\{e^* \partial_{yy}^2 F e\} = \sigma^2 trace[\partial_{yy}^2 F Q_y]$ by virtue of the theorem of section 2, it follows upon taking the expectation of (33) that

$$b_{\hat{x}} \doteq \frac{1}{2} \sigma^2 trace[\partial_{yy}^2 F Q_y] , \tag{34}$$

where we have denoted $E\{\hat{x} - x\}$ by $b_{\hat{x}}$.

In a similar way we find for $\hat{y} = A(\hat{x}) = A(F(y))$:

$$b_{\hat{e}} \doteq -\frac{1}{2} \{ trace[\partial_{xx}^2 A \partial_y F Q_y \partial_y F^*] + \partial_x A trace[\partial_{yy}^2 F Q_y] \} \tag{35}$$

where we have denoted $E\{\hat{e}\}$ by $b_{\hat{e}}$.

Formulae (34) and (35) describe the bias in the least-squares parameter estimator and least-squares residual vector respectively. We know that if the map $A(.)$ is linear both the biases $b_{\hat{x}}$ and $b_{\hat{e}}$ vanish. These biases are however unequal to zero if the map $A(.)$ is nonlinear. Expression (34) and (35) can therefore be considered suitable measures of nonlinearity for the least-squares inverse problem. Moreover, the biases (34) and (35) can be tested on their significance in a way analogous to the well-known procedures of the theory of hypothesis testing.

In order to compute the biases $b_{\hat{x}}$ and $b_{\hat{e}}$ we need to know the partial derivatives of the map $F(.)$. These derivatives are found in the following way. We start from the orthogonality condition (31a):

$$0 = \partial_x A(\hat{x})^* Q_y^{-1} e(\hat{x}) . \tag{36}$$

A Taylor expansion at x gives the following expansion in $\Delta \underline{x} = \hat{\underline{x}} - x$:

$$0 = \partial_x A(x)^* Q_y^{-1} \underline{e} + [\partial_{xx}^2 A(x) Q_y^{-1} \underline{e} - \partial_x A(x)^* Q_y^{-1} \partial_x A(x)] \Delta \underline{x} + \cdots \tag{37}$$

We now substitute our first expansion in $\underline{e}$, (33), into the above expansion. The result is a new expansion in $\underline{e}$:

$$0 = [\partial_x A(x)^* Q_y^{-1} - \partial_x A(x)^* Q_y^{-1} \partial_x A(x) \partial_y F] \underline{e} + \cdots \tag{38}$$

This expansion is identical to zero for all $\underline{e}$. Hence we may collect terms of the same order and set them to zero. For the first order term this gives:

$$\partial_y F = Q(x) \partial_x A(x)^* Q_y^{-1} . \tag{39}$$

Note that when the map $A(.)$ is linear, substitution of (39) into (33) gives indeed the *linear* least-squares estimator of x.

The second order partial derivatives of $F(.)$ are obtained in a similar way. If we substitute the results obtained for $\partial_y F$ and $\partial_{yy}^2 F$ into (34) and (35) we finally get

$$
\begin{array}{lll}
a) & b_{\hat{x}} \doteq & Q(x) \partial_x A(x)^* Q_y^{-1} b \\
b) & b_{\hat{e}} \doteq & [I - \partial_x A(x) Q(x) \partial_x A(x)^* Q_y^{-1}] b , \quad with \\
c) & b = & -\frac{1}{2} \sigma^2 trace[\partial_{xx}^2 A(x) Q(x)]
\end{array}
\tag{40}
$$

Note the resemblance between the expression for b in (40) and the espression (15) for $E\{\underline{R}_2\}$. Also note that the biases $b_{\hat{x}}$ and $b_{\hat{e}}$ can be computed from the vector b just like in the *linear* least-squares case the estimators $\hat{\underline{x}}$ and $\hat{\underline{e}}$ are computed from the data vector $\underline{y}$. Hence, with an available standard least-squares software package the evaluation of $b_{\hat{x}}$ and $b_{\hat{e}}$, and thus the evaluation of the effect of nonlinearity on the nonlinear least-squares estimators, can become rather straightforward.

With our result (40) it now also becomes possible to connect the bias measures of nonlinearity with the geometric measures of nonlinearity of the previous section. Let us first consider $b_{\hat{x}}$. In the expression for $b_{\hat{x}}$ we recognize $\partial_x A(x)^* Q_y^{-1} b$ as the tangential components of b. This shows that $b_{\hat{x}}$ must be related to our geometric measure of nonlinearity of the parameter curves, refer to (28) or (29). This becomes even clearer if we write $b_{\hat{x}}$ as

$$b_{\hat{x}} \doteq [\partial_x A(x)]^- P_{\partial_x A} b \,, \tag{41}$$

where $[\partial_x A(x)]^-$ is an arbitrary inverse of $\partial_x A(x)$ and $P_{\partial_x A}$ is the orthogonal projector onto $R(\partial_x A(x))$. If we let t_α, $\alpha = 1, \ldots, n$, be an orthonormal basis of $R(\partial_x A(x))$ we can write $P_{\partial_x A}$ as

$$P_{\partial_x A} = \sum_{\alpha=1}^{n} t_\alpha t_\alpha^* Q_y^{-1} \,, \tag{42}$$

Substitution into (41) gives with (40c),

$$b_{\hat{x}} \doteq -\frac{1}{2}\sigma^2 [\partial_x A(x)]^- \sum_{\alpha=1}^{n} t_\alpha \, trace[t_\alpha^* Q_y^{-1} \partial_{xx}^2 A(x) Q(x)]$$

or

$$\boxed{b_{\hat{x}} \doteq -\frac{1}{2}\sigma^2 [\partial_x A(x)]^- \sum_{\alpha=1}^{n} t_\alpha \sum_{\beta=1}^{n} \gamma_{t_\alpha}^\beta} \tag{43}$$

where $\gamma_{t_\alpha}^\beta, \beta = 1, \ldots, n$ are the extreme values of the ratio (28). This result shows how the nonlinearity of the parameter curves determines the bias in the nonlinear least-squares parameter estimators.

Following a completely analogous derivation as above one can show [Teunissen, 1985] that the bias in the least-squares residual vector can be expressed in terms of the principal normal curvature as

$$\boxed{b_{\hat{e}} \doteq -\frac{1}{2}\sigma^2 \sum_{i=1}^{m-n} n_i \sum_{j=1}^{n} k_{n_i}^j} \tag{44}$$

This result shows how the local geometry of the manifold $A(x)$ determines $b_{\hat{e}}$. Note that equation (44) clearly shows that the bias $b_{\hat{e}}$ is invariant under a change of variables.

5 DIAGNOSING NONLINEARITY

With the results of the previous sections we are now in a position to propose a strategy for diagnosing the amount of nonlinearity for the problem of nonlinear least-squares inversion. We suggest a two step procedure, see figure 3.

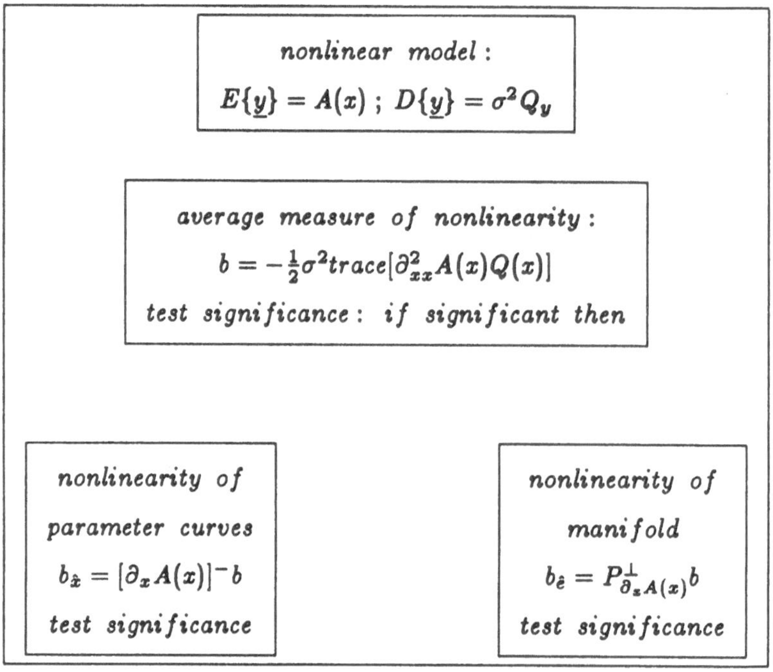

$$\boxed{\begin{array}{c} nonlinear\ model: \\[4pt] E\{\underline{y}\} = A(x)\ ;\ D\{\underline{y}\} = \sigma^2 Q_y \end{array}}$$

$$\boxed{\begin{array}{c} average\ measure\ of\ nonlinearity: \\[4pt] b = -\tfrac{1}{2}\sigma^2 trace[\partial^2_{xx} A(x) Q(x)] \\[4pt] test\ significance:\ if\ significant\ then \end{array}}$$

$$\boxed{\begin{array}{c} nonlinearity\ of \\ parameter\ curves \\[4pt] b_{\hat{x}} = [\partial_x A(x)]^- b \\[4pt] test\ significance \end{array}} \qquad \boxed{\begin{array}{c} nonlinearity\ of \\ manifold \\[4pt] b_{\hat{e}} = P^{\perp}_{\partial_x A(x)} b \\[4pt] test\ significance \end{array}}$$

Figure 3: Diagnosing nonlinearity of least-squares inversion.

We suggest that one first computes the vector

$$b = -\frac{1}{2}\sigma^2 trace[\partial^2_{xx} A(x) Q(x)] . \tag{45}$$

For those problems where it is difficult to compute the vector of Hessian matrices, $\partial^2_{xx} A(x)$, analytically, a numerical finite difference scheme may be used. The significance of the average measure of nonlinearity b may be tested elementwise $b^i, i = 1, \ldots, m$ or for the length of the vector b, $\| b \|_{Q_y}$. If $\| b \|_{Q_y}$ is insignificant, then also $\| b_{\hat{x}} \|_{Q(x)}$ and $\| b_{\hat{e}} \|_{Q_y}$ are insignificant. This follows since the following bounds hold

$$\| b_{\hat{x}} \|_{Q(x)} \leq \| b \|_{Q_y} \quad and \quad \| b_{\hat{e}} \|_{Q_y} \leq \| b \|_{Q_y} . \tag{46}$$

The proof of these bounds goes as follows. From (40) it follows that

$$\| b_{\hat{x}} \|_{Q(x)} = \| P_{\partial_x A} b \|_{Q_y} \quad and \quad \| b_{\hat{e}} \|_{Q_y} = \| P^{\perp}_{\partial_x A} b \|_{Q_y} , \tag{47}$$

where $P_{\partial_x A}$ and $P_{\partial_x A}^\perp$ are the orthogonal projectors that project onto $R(\partial_x A(x))$ and $R(\partial_x A(x))^\perp$ respectively. Using the *Pythagorean theorem* we get

$$\| b \|_{Q_y}^2 = \| b_{\hat{x}} \|_{Q(x)}^2 + \| b_{\hat{e}} \|_{Q_y}^2 , \tag{48}$$

which proofs (46).

In a somewhat similar way we find with the help of the Cauchy-Schwarz inequality for the individual bias components the upperbounds

$$| b_{\hat{x}}^\alpha | \leq \sigma_{\hat{x}}^\alpha \| b_{\hat{x}} \|_{Q(x)} \leq \sigma_{\hat{x}}^\alpha \| b \|_{Q_y} , \quad \alpha = 1, \ldots, n \tag{49}$$

and

$$| b_{\hat{e}}^i | \leq \sigma_{\hat{e}}^i \| b_{\hat{e}} \|_{Q_y} \leq \sigma_{\hat{e}}^i \| b \|_{Q_y} , \quad i = 1, \ldots, m . \tag{50}$$

If the average measure of nonlinearity b is found to be significant one can go one step further and analyze the nonlinearity of both the parameter curves and the manifold. This is done by computing

$$b_{\hat{x}} = [\partial_x A(x)]^- b \quad and \quad b_{\hat{e}} = P_{\partial_x A(x)}^\perp b , \tag{51}$$

where $[\partial_x A(x)]^-$ is a least-squares inverse of $\partial_x A(x)$. The computation of (51) can be done with a standard least-squares software package. The tests of significance for the effect of nonlinearity on the least-squares parameter estimator and least-squares residual vector can then again be done either elementwise or for the length of the vectors $b_{\hat{x}}$ and $b_{\hat{e}}$. If either $b_{\hat{x}}$ or $b_{\hat{e}}$ is found to be significant one should seriously reconsider whether it is practically justified to apply the theory of linear inference to the nonlinear inversion problem. If one comes to the conclusion that the *nonlinear* probability density functions of the estimators, instead of their linearized versions are needed, then one should follow one of the procedures which were briefly discussed in the introductory section 1.

6 ACKNOWLEDGEMENT

This work was partially funded by the Royal Netherlands Academy of Sciences (KNAW) and the Netherlands Geodetic Commission (NGC). The expert typesetting of this work by Ms. ir. H.M.E. Verhoef using the program LaTeX is greatly appreciated.

7 REFERENCES

Amari S.I., Barndorff-Nielsen O.E., Kass R.E., Lauritzen S.L., Rao C.R. (1987): *Differential geometry in statistical inference*, Inst. of Math. Stat., Vol. 10.

Austin J.W., Leondes C.T. (1981): *Statistically Linearized Estimation of Reentry Trajectories*, IEEE Trans. Aerospace and Electr. Syst., Vol. AE-17, No. 1, pp. 54-61.

Baarda W. (1967): *Statistical Concepts in Geodesy*, NGC, Publications on Geodesy, New Series, Vol. 2, No. 4, Delft.

Bähr H.G. (1985): *Second order effects in the Gauss-Helmert model*, 7th International Symposium on Geodetic Computations, Cracow, Poland.

Bähr H.G. (1988): *A quadratic approach to the non-linear treatment of non-redundant observations*, Manuscripta Geodaetica, Vol. 13, No. 3, pp. 191-197.

Bard Y. (1974): *Nonlinear Parameter Estimation*, Academic Press.

Bierens H.J. (1984): *Robust Methods and Asymptotic Theory in Nonlinear Econometrics*, Lecture notes in Econom. and Math. Syst., Springer-Verlag, Vol. 192.

Blaha G. (1987): *Non-linear parametric least-squares adjustment*, Nova University Oceanographic Center, Scientific Report No. 1.

Bopp H., Krauss H. (1978): *Strenge oder herkömmliche bedingte Ausgleichung mit Unbekannten bei nichtlinearen Bedingungsgleichungen?*, Allgemeine Vermessungs-Nachrichten, Vol. 85, pp. 27-31.

Borre K., Lauritzen S.L. (1989): *Some Geometric Aspects of Adjustment*, In: Festschrift to Torben Krarup, Ed. E. Kejlso et al., Geod. Inst., No. 58, pp. 77-90.

Broyden C.G. (1967): *Quasi-Newton Methods and their Application to Function Minimization*, Mathematics of Computation, Vol. 21, pp. 368-381.

Cauchy A. (1847): *Méthode générale pour la résolution des systèmes d'équations simultanées*, C.R. Acad. Sci. Paris, Vol. 25, pp. 536-538.

Denham W.F., Pines S. (1966): *Sequential Estimation when Measurement Function Nonlinearity is Comparable to Measurement Error*, AIAA Journal, Vol. 4, No. 6, pp. 1071-1076.

Fletcher R., Powell M.J.D. (1963): *A Rapidly Convergent Descent Method For Minimization*, The Computer Journal, Vol. 6, pp. 163-168.

Fletcher R., Reeves C.M. (1964): *Function Minimization by Conjugate Gradients*, The Computer Journal, Vol. 7, pp. 149-153.

Goldfeld S.M., Quandt R.E., Trotter H.F. (1966): *Maximization by quadratic hill climbing*, Econometrica, Vol. 34, pp. 541-551.

Grafarend E.W., Lohse P., Schaffrin B. (1989): *Dreidimesionaler Rückwärtsschnitt*, Zeitschrift für Vermessungswesen, Vol. 114, No. 2, pp. 61-67.

Grafarend E.W., Schaffrin B. (1989): *The Geometry of Nonlinear Adjustment - the Planar Trisection Problem*, In: Festschrift to Torben Krarup, Ed. E. Kejlso et al., Geod. Inst., No. 58, pp. 149-172.

Graybill F.A. (1976): *Theory and Application of the Linear Model*, Duxburry Press.

Haggan V., Ozaki T. (1981): *Modelling Nonlinear Random Vibrations Using an Amplitude-Descent AR Time Series Model*, Biometrika, Vol. 68, pp. 189-196.

Heck B. (1988): *The Nonlinear Geodetic Boundary Value Problem in Quadratic Approximation*, Manuscripta Geodaetica, Vol. 13, No. 6, pp. 337-348.

Jennrich R.I. (1969): *Asymptotic Properties of Nonlinear Least Squares Estimators*, The Annals of Mathematical Statistics, Vol. 40, pp. 633-643.

Jeudy L.M.A. (1988): *Generalyzed variance-covariance propagation law formulae and application to explicit least-squares adjustments*, Bull. Geod., Vol. 62, No. 2, pp. 113-124.

Jupp D.L.B., Vozoff K. (1975): *Stable Iterative Methods for the Inversion of Geophysical Data*, Geophys. J.R. Ast. Soc., Vol. 42, pp. 957-976.

Keenan D.H. (1985): *A Tukey Non-Additivity-Type Test for Time Series Nonlinearity*, Biometrika, Vol. 72, pp.39-44.

Kelly R.P., Thompson W.A. (1978): *Some Results on Nonlinear and Constrained Least-Squares*, Manuscripta Geodaetica, Vol. 3, pp. 299-320.

Kennett B.L.N. (1978): *Some Aspects of Nonlinearity in Inversion*, Geophys. J.R. Astr. Soc., Vol. 55, pp. 373-391.

Koch K.R. (1988): *Parameter Estimation and Hypothesis Testing in Linear Models*, Springer Verlag.

Koch K.R. (1989): *Bayesian Inference with Geodetic Applications*, In print.

Krarup T. (1982): *Non-linear adjustment and curvature*, In: Forty Years of Thought, Delft, pp. 145-159.

Krebs V. (1980): *Nichtlineare Filterung*, Oldenbourg Verlag, München.

Kubik K.K. (1967): *Iterative Methoden zur Lösung des nichtlinearen Ausgleichungsproblemes*, Zeitschrift für Vermessungswesen, Vol. 91, No. 6, pp. 214-225.

Kubik K.K. (1968): *On the efficiency of least-squares estimators in non-linear models*, Statistica Neerlandica, Vol. 22, No. 1, pp. 33-36.

Larson R.E., Dressler R.M., Ratner R.S. (1967): *Application of the Extended Kalman Filter to Ballistic Trajectory Estimation*, Stanford Res. Inst.

Levenberg K. (1944): *A method for the solution of certain nonlinear problems in least-squares*, Quart. Appl. Math. Vol. 2, pp. 164-168.

Liang D.F., Christensen G.S. (1975): *Exact and Approximate State Estimation for Non-linear Dynamic Systems*, IIAC, Automatica, Vol. 11, pp. 603-613.

Lines L.R., Treitel S. (1984): *A review of Least-Squares Inversion and its Application to Geophysical Problems*, Geophysical Prospecting, Vol. 32, pp. 159-186.

Loomis L.H., Sternberg S. (1968): *Advanced Calculus*, Addison-Wesley.

Marquardt D.W. (1963): *An algorithm for least-squares estimation of nonlinear parameters*, J. SIAM II, pp. 431-441.

Mehra R.K. (1971): *A Comparison of Several Nonlinear Filters for Reentry vehicle Tracking*, IEEE Trans. Autom. Contr., Vol. AC-16, No. 4, pp. 307-319.

Mood A.M., Graybiull F.A., Boes D.C. (1974): *An Introduction to the Theory of Statistics*, McGraw Hill.

Oldenburg D.W. (1983): *Funnel Functions in Linear and Nonlinear Appraisal*, Journal of Geophysical Research, Vol. 88, No. 39, pp. 7387-7398.

Ortega J.M., Rheinboldt W.C (1970): *Iterative solution of nonlinear equations in several variables*, Academic Press, New York.

Ortega J.M. (1987): *Matrix Theory*, Plenum Press.

Pazman A. (1987): *On Formulas for the Distribution of Nonlinear L.S. Estimates*, Statistics, Vol. 18, No. 1, pp. 3-15.

Pope A. (1972): *Some pitfalls to be avoided in the iterative adjustment of non-linear problems*, Proc. 38th Annual Meeting, American Society of Photogrammetry.

Pope A. (1982): *Two approaches to non-linear least-squares adjustments*, The Canadian Surveyor, Vol. 28, No. 5, pp. 663-669.

Rothman D.H. (1985): *Nonlinear Inversion, Statistical Mechanics and Residual Statics Estimation*, Geophysics, Vol. 50, pp. 2797-2807.

Saito T. (1973): *The non-linear least-squares of condition equations*, Bull. Geod., Vol. 110, pp. 367-395.

Sanso F. (1973): *An Exact Solution of the Roto-Translation Problem*, Photogrammetrica, Vol. 29, pp. 203-216.

Schaffrin B. (1985): *A note on linear prediction within a Gauss-Markoff model linearized with respect to a random approximation*, Proc. First Tampere Sem Linear Models, Univ. Tampere, pp. 285-300.

Schek H.J., Maier Ph. (1976): *Nichtlineare Normalgleichungen zur Bestimmung der Unbekannten und deren Kovarianzmatrix*, ZfV, Vol. 101, No. 4, pp. 140-159.

Schmidt W.H. (1982): *Testing Hypothesis in Nonlinear Regressions*, Math. Operations forsch. Statist., Secr. Statistics, Vol. 13, No. 1, pp. 3-19.

Sorenson H.W. (1977): *Approximate Solutions of the Nonlinear Filtering Problem*, IEEE Descision and Control Conference, pp. 620-625.

Spivak M. (1979): *A Comprehensive Introduction to Differential Geometry*, Vol. II, Publish or Perish, Inc.

Stark E., Mikhail E. (1973): *Least-Squares and Nonlinear Functions*, Photogrammetric Engineering, pp. 405-412.

Tarantola A., Valette B. (1982): *Generalized Nonlinear Inverse Problems Solved Using the Least-Squares Criterion*, Reviews of Geophysics and Space Physics, Vol. 20, No. 2, pp. 219-232.

Tarantola A. (1986): *A Strategy for Nonlinear Elastic Inversion of Seismic Reflection Data*, Geophysics, Vol. 51, No. 10.

Tarantola A., Jorbert G., Trézégnet D., Denelle E. (1987): *The Use of Depth Extrapolation for the Nonlinear Inversion of Seismid Data*, Geophysics.

Tarantola A. (1987): *Inverse Problem Theory*, Methods for Data Fitting and Model Parameter Estimation, Elsevier.

Teunissen P.J.G. (1984): *A note on the use of Gauss' formula in nonlinear geodetic adjustments*, Statistics and Descisions, No. 2, pp. 455-466.

Teunissen P.J.G. (1985): *The geometry of geodetic inverse linear mapping and nonlinear adjustment*, NGC, Publ. on Geodesy, New Series, Vol. 8, No. 1, Delft.

Teunissen P.J.G. (1985a): *Nonlinear adjustment: An introductory discussion and some new results*, In: Proc. SSG 4.56, Workshop Meeting, Ghania, Greece, pp. 10-12.

Teunissen P.J.G. (1987): *The 1D and 2D Symmetric Helmert Transformation: Exact Nonlinear Least-Squares Solutions*, Reports of the Department of Geodesy, Section Mathematical and Physical Geodesy, No. 87.1, Delft.

Teunissen P.J.G. (1988a): *The Nonlinear 2D Symmetric Helmert Transformation: An Exact Nonlinear Least-Squares Solution*, Bull. Geod., Vol. 62, No. 1, pp. 1-16.

Teunissen P.J.G. (1988b): *First and Second Order Moments of Nonlinear Least-Squares Estimators*, will be published in Bull. Geod..

Teunissen P.J.G. Knickmeyer E.H. (1988): *Nonlinearity and Least Squares*, CISM Journal ACSGC, Vol. 42, No. 4, pp. 321-330.

Teunissen P.J.G. (1989): *A Note on the Bias in the Symmetric Helmert Transformation*, In: Festschrift to Torben Krarup, Ed. E. Kejlso et al., Geod. Inst., No. 58, pp. 335-342.

Tong H. (1986): *Nonlinear Time Series Models of Regularly Sampled Data*, Bernoulli Soc. for Math. Stat. and Prob., Tashkent.

Wiggens R.A. (1972): *The General Linear Inverse Problem: Implications of Surface Waves and Free Oscillations for Earth Structure*, Rev. Geophys. Space Phys., Vol. 10, pp. 251-285.

Williams C.A., Richardson R.M. (1988): *A Nonlinear Least-Squares Inverse Analysis of Strike-Slip Faulting with Application to the San Andreas Fault*, Geophysical Research Letters, Vol. 15, No. 11, pp. 1211-1214.

Wolf H. (1961): *Das Fehlerfortpflanzungsgesetz mit Gliedern II Ordnung*, ZfV, Vol. 86, pp. 86-88.